NEW PERSPECTIVES ON CYBERNETICS

SYNTHESE LIBRARY

STUDIES IN EPISTEMOLOGY,

LOGIC, METHODOLOGY, AND PHILOSOPHY OF SCIENCE

VOLUME 220

NEW PERSPECTIVES ON CYBERNETICS

Self-Organization, Autonomy and Connectionism

Edited by

GERTRUDIS VAN DE VIJVER
University of Ghent, Belgium

KLUWER ACADEMIC PUBLISHERS
DORDRECHT / BOSTON / LONDON

Library of Congress Cataloging-in-Publication Data

New perspectives on cybernetics : self-organization, autonomy, and connectionism / edited by Gertrudis van de Vijver.
p. cm. -- (Synthese library ; v. 220)
Includes bibliographical references and index.

1. Cybernetics. 2. Self-organizing systems. 3. Automatic control. I. Vijver, Gertrudis van de. II. Series.
Q310.N48 1992
003'.5--dc20 91-34827

ISBN 978-90-481-4107-4

Published by Kluwer Academic Publishers,
P.O. Box 17, 3300 AA Dordrecht, The Netherlands.

Kluwer Academic Publishers incorporates
the publishing programmes of
D. Reidel, Martinus Nijhoff, Dr W. Junk and MTP Press.

Sold and distributed in the U.S.A. and Canada
by Kluwer Academic Publishers,
101 Philip Drive, Norwell, MA 02061, U.S.A.

In all other countries, sold and distributed
by Kluwer Academic Publishers Group,
P.O. Box 322, 3300 AH Dordrecht, The Netherlands.

Printed on acid-free paper

CONTENTS

PREFACE

*Gertrudis Van de Vijver**

Seminar of Logic and Epistemology
University of Ghent

Before being classified under the fashionable denominators of complexity and chaos, self-organization and autonomy were intensely inquired into in the cybernetic tradition. Despite all rejections that cybernetics has gone through in the second half of this century, today its importance is more and more recognized. Its decisive influence for connectionist theories, autopoietic and constructivist theories, for different forms of applied or experimental epistemology, is being more and more understood and generally accepted.

It is mainly due to the success of connectionist models that we observe today a revival of interest for cybernetics. The 1943 article by McCulloch and Pitts is evidently a founding article. Cybernetics has however a much broader interest than the one linked to technical-mathematical details relevant to the construction of networks. For instance, the evolution from first to second order cybernetics, the ways of approaching biological and cognitive phenomena in the latter and the limits that were formulated there, are particularly meaningful to understand current developments and divergences in connectionism. A nuanced picture of cybernetic's history and its present state is therefore clearly epistemologically essential.

The major differences between first and second order cybernetics are traced out in the Introduction by Gordon Pask. We will therefore limit ourselves to a brief sketch of some elements of the history of cybernetics — a history full of misunderstandings, mistaken interpretations, denials and rejections — which are of importance for the studies that are nowadays to be situated under the denominators of self-organization and complexity.

First order cybernetics, starting at the beginning of the forties with the work of Wiener, Rosenblueth, Bigelow, McCulloch, Pitts, and many others, was definitely an interdisciplinary project. It was, as a consequence, marked by a great diversity in aims and by quite divergent views on how cognitive, biological and teleological phenomena had to be studied and explained. Although it is generally described, rather monolithically, as the *science* of the control and communication in the animal and the machine, its status and meaning as a scientific discipline were not and are not that straightforward.

G. Van de Vijver (ed.), New Perspectives on Cybernetics, 1–9.

Is cybernetics a *science* of feedback-mechanisms or of information ? Does it give expression to the idea of a neo-mechanicism, an 'experimental philosophy' relevant to very different fields of research ? Is it rather a *methodology of action*, a strategy, an art ? Or does it have in the first place a technological value ?

As Gordon Pask states in the Introduction to this volume, the interpretation of the meaning of cybernetics depends on the philosophical orientation one has with regard to science. It is indeed not at all evident that experiment, technology and art ought to be seen in all cases as distinct.

These distinctions however did play a role in the reception of cybernetics in the scientific community.

The technological side of the cybernetical project never posed many problems. Cybernetics was, certainly after 1953, mainly identified with this aspect and its importance was accepted as such. The following statement by McCulloch, in 1961, is characteristic in this respect: "In English medicine cybernetics is still a dirty word, but in their industry it has been washed in the holy water of filthy lucre." (McCulloch, 1965, pp. 221-222).

The reception of what we called the experimental side of the cybernetic project was and is, however, much more delicate. The importance and diversity of the cybernetical research, from its beginning till second order cybernetics, makes it quite incomprehensible that it has been rejected for such a long time on the basis of a monolithic and necessarily historically mistaken picture, a picture which reduces its value to a mere technological one. Questions about the scientific character, about the objectivity of cybernetics remain vivid up to this day, leading to surprised reactions in seeing cybernetic research continuing, and to quick answers such as: "Cybernetics ? Cybernetics is dead!".

Was it the aim to understand in a scientific way teleological phenomena that caused the great stir being made in the scientific community ? Was this the reason for the all too passionate and general rejection that lasted for more than 40 years ? Was it the general interdisciplinary project which was in the first place represented by Norbert Wiener ?

That cybernetics was generally subjected to criticism and rejection, has indeed, according to us, to do with the aim of understanding, within a scientific context, teleological phenomena in a broad sense. In this context, we can refer to the reactions on the founding article "Behaviour, Purpose, Teleology". However, we do not want to deal with this aspect here; we will briefly indicate what are the major shifts in the view on teleology from first cybernetics to the new cybernetics.

In first cybernetics, as we can see in the above mentioned article, teleology was essentially connected with the possibility of control, and necessitated the presence of a well-defined goal-object and a completely specified system. External observation, a classical view on objectivity in which subject and object are separable, were some of the epistemological features of first order cybernetics. However, the fact of stressing the necessity of control had not only as a consequence that technological aims were quite naturally pursued in the first place, it also implied that some forms of purposiveness could not be accounted for. The first cybernetics' interpretation implied essential simplifications on the level of the *meaning* of goal-directedness.

The teleological problem as Kant had formulated it in connection with biological organizations, and that we can describe to-day in terms of self-organization and autonomy, necessitates another approach than the one of first cybernetics in terms of control. The paradox that Kant had linked to teleology (or to internal purposive forms or natural purposes) is related to the fact that a purposive system has to move or develop towards a purpose before that purpose is present, apparently even before a purpose can be conceived of. A genuine purposive system does not only possess a representation of the purpose towards which it is moving, it also has to construct that representation itself.

Along the main lines of the cybernetical project, it had to be possible to understand and explain how certain kinds of systems — artificial and biological — can develop certain goals, how they come to behave in a purposive manner. The necessity for a theory of goals, i.e. a theory of meaning, which explicates how boundaries and goals are created, not only *for* a system, but also *by* a system, clearly presented itself after first cybernetics. Moreover, a theory of how observation plays a central role in the description and explanation of teleological phenomena had to be developed; external observation turned out to be inadequate for complex systems.

First order cybernetics is a theory of observed systems. Second order cybernetics is a theory of observing systems, concerned with problems of biological organization and with the paradoxes of self-reference and self-organization that can be linked to it.

The 'post-cybernetics' or 'new cybernetics' that we are confronted with today is fundamentally to be seen as a continuation of second-order cybernetics.

The original aim of cybernetics is retained in that the artificial system is still considered to be an important means in the clarification and explanation of natural systems and phenomena. However, through the complexification of the artificial system, the traditional borderlines between natural and artificial, which were hardly questioned during the first cybernetics, are becoming more and more problematic. With regard to the latter we can even state that the artificial system can be seen as a very specific source of experience and knowledge. The effect of this on human and 'hard' sciences has yet to be explored in all its riches.

The complexification of the artefact obliges us to reconsider thoroughly concepts like observation and objectivity. Consequences on the epistemological and sociological level, as well as consequences for the relation between theory and praxis, have to be made more explicit. The evolution out of first cybernetics confronted us on the one hand with the impossibility to control autonomous systems and on the other hand with the task to precisely delineate the limitations of knowing and observing self-organizing systems. Whereas control was one of the key-concepts of first cybernetics, in the post-cybernetic theories however, interaction has taken its place. The role of multiple soft 'constraints', as such not sufficient, neither necessary, has to be taken into account to deal adequately with autonomous and self-organizing systems.

This collection, containing epistemological and sociological contributions about autonomy and self-organization in a new cybernetic perspective, is in several respects quite significant. The theories which are under consideration in this volume, were, at least partially, developed out of the impasses of first order cybernetics. Some of these belong explicitly to the second order cybernetics as it took form starting with the publications about self-organization by Von Foerster and Ashby. The current constructivist school of thought has been heavily inspired by Von Foerster's epistemological writings. Other theories, in particular the theory of autopoietic systems by Maturana and Varela, deal with the autonomy of biological organization with the organizational closure of autonomous systems, and the knowability of these. Though the latter do not wish to be classified among the second order cyberneticians, their ideas apparently took shape starting from the same problems and impasses. In terms of openness/closure, adaptation/mimesis, construction/enaction, representation/adaptation/enaction, etc., the consequences of an alternative view on objectivity are here equally traced out.

The articles have been arranged as follows:

i) The first part of this book consists of the search for formal models which give expression to self-organizing and autonomous properties. It inquires into the differences that are found between natural and artificial systems.
ii) In the second part, the epistemological status of the new cybernetics as well as of the theories of autopoiesis and constructivism, is inquired into. In addition,

an alternative minimalistic view on epistemology aims at making the specific status of experimental epistemology explicit.
iii) The implications of those theories for sociology and political theory form the third part of this book.

i) Self-organization and complexity

First we shall discuss some articles containing a formal elaboration of concepts which were crucial in second order cybernetics.

In *Ends and meaning in machine-like systems*, Henri Atlan is concerned with one of the most complex problems of the second order cyberneticians: the problem of meaning. For them, the emergence of meaning was tied up with the finality and functionality of systems. Atlan's interpretation of both finality and functionality shows, however, a turnabout with regard to second order cybernetics: for him, meaning precedes finality. He discusses the possibility of mechanisms whereby meaning can emerge in natural systems without having to define a priori the ends the systems are pursuing. Initially, meaning is created as a classification criterion; later, or at the very utmost simultaneously in relation to it, the function is defined as the classification according to that criterion. The relation between ends and meaning is illustrated through the formalism of the Boolean networks.

The text of Stanley Salthe, *Hierarchical non-equilibrium self-organization as the new post-cybernetic perspective*, criticizes the concepts of hierarchy and organization, starting from first order cybernetics — cybernetics from the perspective of equilibrium. Moreover, it has the intention of laying the conceptual bases for the dynamics of natural, self-organizing systems on the basis of what he calls a post-cybernetic perspective of non-linearity, of non-equilibrium. These dynamics not only apply to the functioning, but also to the development and the evolution of that kind of systems.

In a formal elaboration by Elie Bernard-Weil, the relation between theory and practice, openness and closure, between observing and being observed, is explicitly under discussion. In *A priori and a posteriori in cognitive praxis: the model for the regulation of agonistic antagonistic couples*, the impossibility of a theory of cognition as separable from a praxis is analyzed, specifically in connection with medical discourse. Rather than considering it as separated from praxis, as in most of the 'unilateral strategies', cognition should be seen as an epigenetic development out of praxis. 'Bilateral strategies' are based upon this idea and aim at avoiding some of the 'perverse' effects of the unilateral approaches.

Francis Heylighen's article, *Non-rational cognitive processes as changes of distinctions*, is to be situated in continuation of Spencer-Brown's logic of distinctions, which was studied thoroughly in the School of Santiago by Varela and Maturana. On the basis of the definition of rational processes as distinction-preserving, a classification is worked out which should make it possible to model non-rational processes as incompletely distinction-conserving. Rational cognitive processes are processes constrained by an external system of rules, and these constraints are represented by means of the conservation of distinctions.

ii) Epistemological Issues

In *Self-organization and autonomy in a post-cybernetic perspective: epistemological issues*, Fernand Vandamme gives an illustration of a minimalistic approach in epistemology. Knowledge, seen as the structuration of a specific domain, is at best a minimal structuration of that domain, and this in view of a maximal adaptation with regard to ever changing circumstances. The minimalism means on the one side that striving for a complete control does not make sense, and on the other side it refers to the fact that all knowledge is to be related to concrete circumstances and to the purposes which are postulated by the observing system. The notion of 'being sufficient' is essential in this minimalism. As a matter of fact, the interpretation of self-organizing systems as systems which are able to eliminate and generate structures, finds its basis in the combination of 'sufficiency' and 'possibility'.

Warren Scott McCulloch was one of the leading figures during the first cybernetics. In his work some views are to be found which clearly go beyond these first ambitions and assumptions, not only on account of the epistemological interpretation, but also because of the role he played in the history of connectionism. The following two articles are respectively treating one of those aspects.

In *The experimental epistemology of W.S. McCulloch: A minimalistic interpretation* by Gertrudis Van de Vijver, we find on the one side the delineation of the specific interpretation which McCulloch gave to the experimental epistemology; on the other side we find a delineation of a minimalistic interpretation of experimental epistemology, going beyond McCulloch's rather reductionist interpretation, but always in line with his particular view on 'experimentalism'. Epistemology, in as far as it has the tendency to be experimental, does not serve as a support for reductionism, but is rather directed at the delineation of concrete limits within which a general epistemology can be developed.

Daniel Andler, in *From paleo to neo-connectionism*, questions the pertinency of the always upheld idea of a difference between two forms of connectionism:

paleo- and neo-connectionism. On the basis of two important articles of McCulloch (1943 and 1947), he shows that a clear distinction of embodied logic and embodied computation permits to bring the relation between current connectionism in all its different forms and 'old' connectionism to more nuanced proportions. From Turing's definition of calculability and McCulloch's psychological justification of it onward, this distinction has insufficiently been made explicit up to now. Central in the exposition of D. Andler is the place of representation. Starting from the meaning of this concept in classical cognitivism, he shows in which way this problem did not pose itself for McCulloch, and in which way the success of current connectionism is related to the introduction of a representational level. The differences between the connectionism of the Parallel Distributed Processing and the approach of the Attractor Neural Networks of Amit are positioned with regard to this.

An explicit criticizing of second order cybernetics can be found in the text by Pierre Livet, *Second cybernetics: a double strategy for representing cognition.* The particularly seductive but much too loose character of the formal models in these cybernetics is taken to task by him. On the basis of Von Foerster's criticism of first cybernetics and his interpretation of memory, Livet inquires into the particular view on cognition in the cybernetics of second order. The concept of 'inverted reductionism' is meant to give expression to the specificity and the inadequacies of this view. The topicality of the second cybernetics, via its influence on current connectionism, makes this criticism all the more pertinent.

The next three studies can be seen as critical reflections and further elaborations of the autopoietic theory. They are clearly much more than just sterile repetitions or applications of it. In this theory, biological systems are considered as autopoietic systems, i.e. as systems which are self-producing and self-preserving networks of relations between processes and components.

In his comparison between Latour and Maturana in *The bringing forth of Dialogue: Latour versus Maturana*, Timon Meynen starts from the question of objectivity and the role that can be played by scientific explanation in a 'constructivist' context. He makes explicit some deficiencies with Maturana and shows in which way the infrareflexivity of Latour can counter these.

The article of Arno Goudsmit, *A one-sided boundary: on the limits of knowing organizational closure*, delineates the limitations in observing and knowing self-organizing systems, i.e. the impossibility of controlling autonomous systems we mentioned above. On the basis of Maturana's distinction between openness and closure, he develops meaningful and very nuanced concepts which make us aware of the impossibility to observe the internal states of an autopoietic system in an empirical way. The 'one-sided boundary' is used as a metaphor to give expression to that particular form of inaccessibility. This taking up again of certain phenomenological insights, especially those of Merleau-Ponty, is quite crucial in this context and can be called characteristic for the way in which new cybernetic theories are developed nowadays.

What Paula Burghgraeve states in *Mechanistic explanations and structure-determined systems. Maturana and the human sciences* quite clearly concurs with the points of view and conclusions of Arno Goudsmit, although her starting-point, the interpretation of the mechanistic explanation of Maturana, is different. This article can be of some help to understand the difference between structure-determined systems and systems which interact on the basis of instructive interactions. It also points out which are the consequences of the postulates of the autopoietic theory within the domain of explanation.

iii) Sociological Issues

We have already mentioned that the problem of meaning proved to be central in the cybernetics of second order. This quite naturally led to problems of an ethical and political nature. Once again theory and praxis are seen as inseparably intertwined. Gordon Pask deals with this aspect, starting from the difference in truth as correspondence, coherence and consensus in *Correspondence, consensus, coherence and the rape of democracy*. This discussion has become more topical than ever within the framework of democracy.

The text of Eric Rosseel, *Writers of the lost I: second-order self-observation and absolute writership*, describes and inquires into the psychic consequences of an excessive self-observation. It is quite obvious that this is in line with second order cybernetics. Starting from the problems of the observer, one almost naturally arrives at those of self-observation, and so even, as Rosseel puts it, at 'self-observation of second order'. Within the framework of what we would call a radical constructivist approach, where objects are seen as 'pretensions' (as something which is never given in advance, but which is on the contrary always 'pre-tended'), he describes in which way self-observation can lead to self-destruction, in which way self-observation is changed to its own detriment into second order self-observation. Herewith he does not only concur with the psychic experience which has been so often described in literature and poetry, but he also points out in which way the ability to act is connected to this.

Ghent, September 1991

* Senior Research Assistant Belgian National Fund for Scientific Research

References

Ashby, W.R., 1962, Principles of the self-organizing system, in H. Von Foerster and G.W. Zopf (eds.), *Int. tracts in computer science and technology and their application*, vol. 9: Principles of self-organization, Pergamon Press, Oxford, pp. 255-278.

Maturana, H.; Varela, F.J., 1980, *Autopoiesis and cognition. The realization of the living*, D. Reidel Pub. Co., Dordrecht, Boston Stud. in the Phil. of Science, vol. 42, pp. 140.

McCulloch, W.S., 1965 (1961), Where is Fancy Bred ?, reprinted in *Embodiments of Mind*, The MIT Press, Cambridge, Massachusetts, pp. 72-142.

McCulloch, W.S.; Pitts, W., 1943, A logical calculus of the ideas immanent in nervous activity, *Bulletin of mathematical Biophysics*, vol. 5, pp. 115-133.

Rosenblueth, A.; Wiener, N.; Bigelow, J., 1943, Behaviour, Purpose and Teleology, *Philosophy of Science*, vol. 10, no. 1, pp. 18-24.

Varela, F.J., 1979, *Principles of biological autonomy*, North Holland Elsevier Company, New York (Oxford), The North Holland Series in Gen. Syst. Res., vol. 2, pp. 306.

Von Foerster, H., 1960, On self-organizing systems and their environments, in M.C. Yovits en S. Cameron (eds.), *Int. tracts in computer science and technology and their application*, vol. 2: Self-organizing systems, Pergamon Press, Oxford, pp. 31-51.

INTRODUCTION

DIFFERENT KINDS OF CYBERNETICS*

Gordon Pask

University of Amsterdam
Faculty of Andragological and Educational Studies

Abstract

This paper spells out differences and similarities between the older or classical cybernetics and the recent and very different cybernetics which has been produced as a product of philosophical and scientific evolution. It is argued that considerations such as modal logics (encompassing self and other reference for imperatives, deontic forms and so on) and truth valuations of coherence and distinction are already giving rise to further fundamental developments, some of which are indicated.

1. Introduction

There were two symposia at the meeting on *Communication and Communication. Applied Epistemology*, at the University of Ghent, Belgium. One of the two was titled 'first' (or *old* Cybernetics) and the other 'second' (or *new* Cybernetics). Both symposia were extremely interesting and, to me, at least, very productive. However, due to matters of circumstance, such as travel difficulties and the indisposition of certain participants, there was not enough comparison and contrast between the 'first' and the 'second' varieties of that odd but realistic field of scrutiny and investigation called Cybernetics.

I undertook the task of buttressing the similarities and differences, since I am ancient and involved enough to witness what went on, in person, and if only because my other paper at this congress, in the symposium 'Second Cybernetics', or, as it is often known 'Second-order Cybernetics', is only intelligible if the character of second, or second-order, Cybernetics has been well delineated. Dr. G. Van de Vijver accepted a proposal to render the differences more obvious by writing an additional paper. It is cursory but I hope it will be useful as an informed contribution. Of course, I am not unique in this respect. Both Heinz Von

G. Van de Vijver (ed.), New Perspectives on Cybernetics, 11–31.

Foerster and Gerard de Zeeuw could do a better job. It just happens that I was both there, at Ghent, and was there, some 25 or 35 years ago when the distinctions were in the process of formation.

This paper is an attempt to spell out some salient differences between the *old* and the *new* brands of Cybernetics. It is incomplete, as most papers are, it is biased, as all papers are, towards the particular concerns of their authors; however much these authors may try to suppress their idiosyncratic interests. But it will provide a distinction of sorts if only one caveat is kept firmly in mind.

This caveat is as follows; the *new* incorporates, encompasses, the *old*; but it neither denigrates nor denies the old. All that Ashby, Shannon or Cherry wrote, for example, is correct within its proper and well delineated frame of reference, all that Gabor and McKay said of logons and metrons is correct within their well stipulated arena. The same kind of comment goes for McCulloch and for Wiener, in their, slightly different, frames of reference.

In citing these, or similar sources, chiefly concerned with 'information', I am influenced by an interesting talk given by Prof. Felgett, recently of Reading University, to the London Cybernetic Society. In this talk he contended that Cybernetics is 'the science of information' and is firmly based upon such solid ideas as the maximum capacity of a communication channel. Further, it is a precise discipline spanning many and often quite disparate fields of scientific endeavour. I concur, entirely, with the view that Cybernetics is a precise discipline and is cross disciplinary, but no mere rag bag of useful techniques. On the score of 'information', it might be wise to broaden the definition, but that is probably an uncontentious point. The only real quibble is over the question of whether or not 'Cybernetics is a Science'. Is it perhaps an art, an epistemological scheme, a logic, an ontology, or, as Glanville prefers, 'a philosophy of design' ? That depends, so far as I can see, upon your philosophical orientation with respect to 'science'. For myself, I have never been able to see the distinct cultures of C.P. Snow as divergent; art and science, psychology and aesthetics seem, to me, part of much the same enterprise and they coexist together.

It is, however, necessary to note that even though the *new* brand of Cybernetics does mark a radical paradigm change, it does *not* render such canonical definitions as Wiener's 'science of control and communication in the animal and the machine' or Couffignal's 'l'art d'assurer l'efficacité de l'action' (both popular in the *old* brand) obsolescent, let alone mistaken. It is simply the case that the scope has become more extensive, that the fundamental notions of 'information', of 'system', of 'stability', and so on, need widening if they are to be appropriate to the *new* brand.

2. *Older brand*

In the older Cybernetics it was customary to suppose the existence of a system, essentially specified (and often divided into component subsystems) by an utterly external observer. This system has at any instant of observation (determined by the external observer) a particular state, usually specified by a point in a state space or a phase space. The coordination or the space is also selected, on canonical, common-sensical, or consensual grounds by the observer and the conventional wisdom of the society in which observers exist. A system behaviour is a jointly observable temporal and causal trajectory described by the motion of a state specifying point.

There may be, and frequently are, several a priori independent systems. If so, it is possible for an observer, cognizant of the lot, to combine them by, for example, taking the Cartesian Product of their coordinates and considering a point in the product state space. If such an aggregate exists and if a correlation of behaviour is noticeable, then any directional components reflected, upon further enquiry, may be interpreted as an information flow. Strictly, this involves the putative specification of a communication channel, criteria (observer determined) of relevance and, thus, of 'noise' perturbing the information flow and being essentially and observer determined 'random noise', random meaning (in this case, irrelevant). The observer, if pressed further, may find it necessary to describe the fundamental synchronicity of the interacting systems or subsystems under scrutiny, transmitters and receivers, one at each end of the communication channels. An alternative and perfectly legitimate expedient, used, for instance, in thermodynamics, is to consider an ensemble of a priori independent subsystems which are structurally independent but have differing initial states.

Regarding stability, the state trajectories of a system tend to equilibrium. In the simplest case they converge to one point (or different points from different initial states), more commonly to one (or several, according to initial state) fixed point or limit cycle(s) or even torus(es) of characteristic equilibrial behaviour. All this is, in a rigorous sense, contingent upon describing a potential, as a special summation of the multitude of forces acting upon the state in question, and finding the fixed points, cycles, etc., to be minima of the potential function. It is often more convenient to use the equivalent jargon of 'attractors' so that the state of a system is 'attracted' to an equilibrial behaviour. In any case, once in equilibrial conditions, the system returns to its limit, equilibrial behaviour after small, arbitrary, and random perturbations.

Limit points, cycles and other equilibrial behaviours are often called goals and the system is said to be 'goal-directed'. In a rather weak sense of the term, 'goal-directed' systems have been described as purposive. This conception is by no means trivial, though it may be insufficient, as is amply demonstrated by phenomena such as meta-stability, requiring further coordinates of a state description to be mustered, if the system is to remain in existence or survive.

The entire picture is, however, dominated by observer determination of the system which is observed. This observer is, for example responsible for choosing what *is* an appropriate set of coordinates, which *are* the systems to examine in detail, and so on. There are good grounds for selecting coordinates like temperature, pressure, momentum and other such quantities, especially when speaking of the classical thermodynamic ensemble of independent systems. Again, it seems legitimate to notice that particular horses are legitimately distinct systems, namely, organisms.

It may be good sense, or it may be the merest piffle. That is up to the observer, only. It is piffle, for example, if an observer opts, not too infrequently in real life, to use coordinating variables like 'stress' or 'motivation' (of what, of whom, statistically specified, if so, over which ensemble by what kind of average), when the variables are given values that depend upon 'sound empirical' findings which turn out to be no more than a numerical splodging of responses, 'measured' on a five point scale, to a questionnaire.

On the other hand, if the coordinating variables, choice of system boundaries and so on, are sensible then the stuff of the *old* Cybernetics is entirely congruent with the *new* Cybernetics. The significant findings of the 'first' hold, under appropriate transformation, in the 'second' type. That depends upon the care and wisdom invested by an impartial, external, in a sense numinous outside observer.

3. *New brand*

The 'second' or *new* Cybernetics is characterisable as distinct from the 'first' or *old* version, by the notion of participation. Observation is (in some degree at least) participant observation and this is partial (if needs be, and the bias is announced). In the *new* Cybernetics such things as 'system boundaries' and 'input and output' (supposing the latter to have meaning other than as a convenient and occasionally used expedient) are determined *either* by the system *or* by the observer (insofar as these are properly separable); so are the coordinates describing the system state, or more often, the 'condition', which is a liberalization of 'state'. The observer is a participant, interacting with the system. That, at any rate, is a pivotal point upon which to balance the friendly debate between the Cybernetic fashions and forms.

Let us begin with the notions of *stability* and *system*. It is a reasonable combination insofar as an entity of such erratic behaviour that it exhibited no stability whatever scarcely would be deemed a system. Further, in the *new* cybernetic paradigm which is of immediate concern, the ideas of system and of stability are very intimately related.

Systems, acting dynamically, produce (and incidentally, reproduce) their own boundaries, as structures which are complementary (necessarily so) to their motion and dynamics. They are liable, for all that, to instabilities of chaotic form, where chaos, as commonly interpreted nowadays, is remote from the random. Chaos is a peculiar situation in which the trajectories of a system, taken in the traditional sense, fail to converge as they approach their limit cycles or 'attractors' or 'equilibria'. Instead, they diverge, due to an increase, of indefinite magnitude, in amplification or gain. There may or may not be a random element, such as 'random bombardment', by particles or rays, of 'genetic macromolecules', held responsible for the mutants of an evolutionary process. Even if there were not, chaos, in the given sense, would prevail and do the job of injecting variety, diversity into an otherwise static universe (see also Appendix 1).

In summary, it used to be the case that stability meant something like homeostasis, often formalized to the one or several eigenvalues (maybe cycles or toroidal forms of fixed points) of operators, represented in a state space or phase space. Currently, that meaning of stability, although it rightly persists for special cases, is quite rare. More often we speak of a self-organizing system as 'semi-stable' even if informationally open and evolving. Formally, this can be expressed in terms of several equilibrial conditions or cycles that are 'chaotic'; in particular those chaotic 'attractor' classes (distinct and different equilibrial cycles) associated with regions maintained on a phase-transition boundary (spin glasses were noted by Atlan and it is a happy circumstance, as Jeff Nichols came up with much the same idea about a month ago, in the context of C.T. and the applied epistemology Lp). The point which deserves stress is that when chaos obtains, it is necessary to change the phase space topologically in order to represent the process. The trajectories diverge and differences are amplified with ever increasing gain. It may be said that the system demands a redefinition of its boundaries from the observer (not unlike Ashby's prescient notion of meta-stability, or in structural terms, René Thom). It may also be said, in a special sense, that the system defines its own boundary and is required, under these conditions, to redefine it.

The first *(old)* Cybernetics addressed itself to matters of communication and control. The second Cybernetics still does so. In the second or *new* Cybernetics, however, the overt concern is interaction and I shall illustrate this from a familiar stance (namely, of conversation theory and its protologic Lp, since I was partly responsible for originating them between P-Individuated participants).

There is no doubt that a conversation between participants, a concept exchange, may, often does, lead to sharing of a public concept which may be inscribed, if desired, at the interface of a suitable Lp processor. If such a processor is utilised, then shared concepts may be Lp manipulated in order, for example, to promote or catalyze the conversation. On the other hand, this is not the primary function of a conversation. If participants, say A and B, converse, then they mutually enrich their personal concepts of whatever the conversation is about. Agreements to disagree may give rise to conflict, but that conflict is resolvable insofar as A learns why B is in disagreement and vice versa. In this sense, the primary function of conversation is to deregulate rather than to control, to learn mutualistically of each other and to increase the variety of conceptual repertoires, rather than learning *about* whatever *is* discussed. Of course, if A and B discuss some entity T using some language L then they are quite likely to learn about T. But the main point (as proposed by Shannon and possibly accepted by him in the histories symposium) is that the presence of a common entity is the origin of a representational function with minimal predicate expression of the type Repr(T,L,A,B,...), for we converse in language, the real language of poetry, music and metaphor, question and injunction, as well as assertion and description. It may be noted that something like a Günther logics is mandatory if our aim is to speak of conversation rather than some meta-theoretic notion of it. That would not be necessarily so if participants conversed, for example, by injecting each other with appropriately individuated RNA.

What I have tried to illustrate by way of conversation may be just as well applied to the mutualism of species, the interaction of organisms or symbiosis of host and parasite. It is true of interactions that live, between live creatures. For biology, Maturana coined the term co-ontogeny between autopoietic systems; for immunology Varela, now Varela and Greenberg, have a similar and implemented formulation. Both Maturana and Varela, as well as Ricardo Uribe were well-known visitors, along with Lars Löfgren and myself to Von Foerster's Biological Computer Laboratory (with System Research in the U.K. a positive hotbed of the second Cybernetics, as the O.O.C. programme, led by Gerard de Zeeuw, in Amsterdam at *this* moment).

All of this might have seemed very hazy in the context of the *old* Cybernetics which is paradigmatized by that brilliant thinker, Ross Ashby (also at the Biological Computer Laboratory (B.C.L.) and System's Research (S.R.)). It was obvious only to some, and critically questioned by others, why the first

Cybernetics should dare to subsume a multitude of apparently disparate areas of science or epistemology. In the *new* version, the commonality is much more evident and not so questionable. The interactions between participants in a conversation and organisms and societies now have a unifying theory (but it will be obvious, from the comments upon conversation, that unification does *not* imply *uniformity*). Rather, unity creates variety and the distinction needed to balance the agreement, which in conversation theory is imaged by coherence. Seldom is there complete accord, in man or in society; thankfully agreement is mostly agreement to disagree and often to know why that is so.

Well, just how is this uniformity obtained ? Early in the 1970's I formulated conversational participants, A and B, for example, in terms of P-Individuals or Psychological-social individuals (they could be bits of a cognitive system, people, societies, civilizations, institutions and so on). More or less independently Varela and Maturana came up with autopoiesis, in a primarily biological context and generalized this to the notion of organizational closure. A system which is organizationally closed consists in a collection of productive operators upon some matrix of whatever kind (biological, conceptual, or of other fabric) and *amongst* the products of this system are the productive operators themselves. The system is thus productive and, *incidentally*, reproductive. It turns out that P-Individuals are organizationally closed, that coherent collections of concepts including generalizations and analogies are organizationally closed, that concepts are organizationally closed. They are also informationally open. Further, organisms are organizationally closed, tissues are organizationally closed, cells are organizationally closed and so are many other systems. But, apart from monads and the point repeating 'I' in Edwin Abott's 'Flatland', they are also informationally open. Hence they can, according to the domain that is addressed, converse, interact, or co-ontogenize. Any differences between Maturana and Varela and my own group are, if they exist, differences in notation or nomenclature. For example, Maturana uses, appropriately, the phrase 'structurally open', whereas I use 'informationally open', hence in a position to converse or exchange concepts in order, in both, to obtain a larger closure with, in propitious cases, a preservation of diversity between systems and domains secured by self-organization, as originally stated by Heinz Von Foerster. That attribution, to Von Foerster, is also a reason for the commonality of ideas which characterize the Cybernetics of today and which have all received so much from that source.

It is probably evident, by this stage, that 'information' is given a traditionally unusual meaning. The intended connotation of 'information', at least as it is used in the information-transfer of concept exchange or of co-ontogenic interaction, is coupling of otherwise partly autonomous and independent systems or the partial synchronization of otherwise asynchronous systems. Certain aspects of the transactions can be captured by the combinatorial and statistical indices of syntactic information due to Ashby and Shannon, for instance, or even semantic measures, like those of Bar-Hillel, or more fundamental measures such as the 'logons' and the 'metrons' in the information theory of Gabor and McKay. But the former, coupling and synchronizing idea, is akin to the information transfer occurring in Holt's interpretation of Petri's theory and rightly called Petri information transfer, originally in the context of particle physics but of quite general applicability to all manner of systems.

This latter meaning is, I think, peculiar to the second Cybernetics and the use of the term 'information' in this way, to Bratens 'Dialogic', to much of De Zeeuw and to conversation theory. Neither in conversation theory nor in the much wider context of a second Cybernetics is there a denial or rejection of the other measures and their usefulness. But we are currently dealing with issues like attention directing, the flux of thought and the (different, orthogonal) flux of processes that produce and reproduce concepts, which call for a broader perspective.

Let us say that a system is organizationally closed (autopoietic or P-individuated or socially-individuated, as you wish) or simply that there is a system insofar as there is some fabric (biological, mental, conceptual, societal) upon which productive operators act to give products. *Amongst* these products of the application process, although not exclusively so, are the productive operators themselves. Hence the entire system, including return paths, is not just reproductive in the manner of a printing press, or the parroting of an utterance, but as a productive outcome of application, the process, rightly called life. In my case, since the products are concepts, simpliciter, or concepts that act upon and produce concepts, themselves capable of application to produce products, it is conceptual life.

Of course, these products are neither necessarily nor usually the *only* products of such a viable activity. Although the fundamental, but incidental, *reproduction* ensures the existence and persistence of a boundary or distinction (be it a cell barrier or discrimination of cultures) and although certain exclusion principles do apply (that is, for example, that whilst many operations may lead, by different paths, to one product, no productive operation can, without overtly stated parameter variation, give rise to distinct products) it is still the case that other products may be produced. What of them ? They are not forbidden, except within the formal scheme. They may, and do, exist. They are, however, held distinct, as distinct concepts or even participants. In the latter case, they are and, in the former may be, interactive in what I call a conversation between the participants.

If that is so, then I call the resulting system both organizationally closed and informationally open (using the term 'information' in, roughly, the sense of Carl Adam Petri, to mean the becoming of a coupling between otherwise independent systems, or, equisignificantly, the becoming of a synchronization between hitherto asynchronous systems). Maturana, at least, prefers the phrase 'organizationally closed and structurally open' and he is right, at least in the context of biology. Further, he uses the happily chosen word 'co-ontogeny' to give the meaning of what I intend by 'conversation', namely, a coupling together into a larger closure with the preservation of an open quality. Systems which might, otherwise, have been closed are brought into interaction.

It is, within limits, possible to express these *new* fangled ideas in terms of the *old* nomenclature; the state space; the state trajectory, and so on. The difficulties are, first, to choose canonical variables in terms of which to describe, say, a growing and possibly differentiating cell, tissue, or organism; culture, civilization, person or participant in a conversational exchange; next to deal with the exceptionally common case in which the state trajectories, so represented, tend towards 'attractors' or 'equilibrial cycles', or whatever, that have the superficially perverse character of 'chaos'; that is, as the 'attractor' or 'equilibrium' is approached, the trajectories fail to do the classical trick of converging. They diverge instead. How does one deal with 'chaos', in the dynamic sense or in its structural complement, the 'catastrophes' of Rene Thom's 'Catastrophe Theory'. Well, you bend, topologically deform, the state space and maybe as in classical meta-stability, increase its dimension a bit.

Be clear about this matter. In the conventional representation 'normal' process trajectories form distinctions, a bounding carapace which is complementary to the processes, if they are properly described. So do 'chaotic' attractors and the processes generating them, but the topological husks, their carapaces, their boundaries, the distinctions that they generate, require a different connectivity, topology, often, a higher topological dimension to find expression. That, perhaps, is why they are popularly known as 'chaotic', which is not a bad term but means an ever increasing amplification or 'gain' as the attractor in question is in the immediate vicinity of the state representing point.

Ideas like organizational closure originated in Von Foerster's formulation and development of self-organization (similar work being done by De Zeeuw in Andragology and Braten, in social science). It is no accident that Maturana, Varela, and I (along with Günther, Ashby, Löfgren, and others) were at Von

Foerster's B.C.L.; that Glanville, Robinson, and Uribe were under his influence, or that Bradburne, Martyns, and Navarro, amongst todays exponents, keep in close contact with an ongoing stream of thought. The original idea was phrased in a trickily classical manner, as befits a trickily unclassical, but very wise, very pure, physicist. The very traditional presentation is as follows.

Von Foerster stated, in the idiom (really) of a physicist, that organization viewed as a quality, must have a calculable, ratio, form (like the ratio measures of heat engine efficiency). An obvious choice for this quantity is called 'redundancy', designated R, and its value is given by R=1-H/H(max)=H(max)-H/H(max) for the maximum entropic information possible at a given moment for a system. The system is self-organizing if and only if the rate of change of redundancy is greater than 0, that is dR/dt > 0. An innocent sounding condition, until you look at it rather deeply.

For example, if the system learns, even if it adapts, without changing its maximum complexity, namely H(max), then it will cease to be self-organizing. Such a change as learning, producing a decrease of H, must be compensated by something like attention change, increase in H(max) (which, note, alters the coordinates in which the system is specified, at least insofar as its fundamental degrees of freedom are involved). Conversely, imagine that H is constant, but H(max) changes; what does it entail ? Quite a lot, for how *do* you increase the complexity, H(max), *without* changing H. It is possible, surely enough, but it is difficult and requires the services of a curiously internal coupling or balancing between variations in H and variations in H(max) that respects the existing order and, in effect, specifies the system boundaries, a 'demon' as Von Foerster put it, jokingly, in the spirit of the energetic demon evoked by Maxwell. Any loopholes in this prescient formulation were patched by an augmenting paper, at the Bionics conference in Dayton, Ohio, on molecular ethology. Here, Von Foerster introduces a further condition relating the rate of change of redundancy (or some such form) to the rate of change of a quantity which could be energy (but could be otherwise interpreted, as it should be if the rate derivative is given, as it commonly is nowadays, a less familiar but wider connotation than it has in the standard differential) It was heartening and, I believe entirely veridical, to hear Lineret, of Paris, at the *old* type symposium date the genesis of the *new* paradigm to the work of Von Foerster and the foundation of the Biological Computer Laboratory, the B.C.L.

The emergence of the *new* paradigm was greatly accelerated by the work of C.A. Petri and its interpretation by Anatol Holt, given for example, at one of Gregory Bateson's symposia at Borg Wartgenstein; establishing the ideas of concurrency, of conflict and of conflict resolution. The *new* paradigm gained creditable substance, at this point, sufficient to support a good deal of the more or less technical work which, though valuably ongoing, needed it to furnish its firm bedrock. Again, let me stress that the *new*, whilst it modifies the epistemology, logic and ontology of Cybernetics, does not (except in the near silly cases which I tried to parody) deny the *old*, neither were the *old* practitioners entirely ignorant of it. McCulloch, for example, was as much concerned (bound up in, pardon the pun) knots and braids, or in $n > 2$-adic, irreducible, logics (stuff of the new) as in McCulloch and Pitts networks. Many others were in cahoots, Bigelow, Rosenblueth, Wiener on his visits to Caianiello in Napoli, Sanna, in Genova, to list a few of them.

There are many aspects to the *new* paradigm. I am stressing a few, namely, the logics required in order to deal with participant observation, organizational closure, the generation of distinctions, conversation, or co-ontogeny forming larger closures, this very conversation guaranteeing an enrichment of the diversity of the participants. These are modal logics, most or all, based upon logics of distinction and coherence, distinctions of many sorts, coherencies of many sorts, self and other reference and validity in some but not always all possible domains of interpretation.

The second order of *new* Cybernetics is greatly concerned with distinction, notably with system boundaries and distinctions that are system created. P-Individuation and organizational closure are cases in point. On the whole, first-order Cybernetics consigned these matters to a meta-theory, as in the transitional work of Beer, or neglected them. But Beer is nowadays greatly concerned with self-made distinction, as are Flores and Winograd in their recent work.

It is, of course, up to an observer (an external observer, that is), to distinguish between systems such as the participants A and B in any coherent or (loosely speaking) consistent way. Hence, distinctions like those established by organizational closure are amongst many possibilities. A and B could be demarcated, for example, in terms of skin, face, race, immune system, blood type and so on. But, in a certain sense, the distinction wrought as the structural complement to the process which constitutes the system is canonical. It is canonical for an observer who is not an external observer but is a participant.

It is fair to say that Cybernetics nowadays is primarily a science, or as Glanville puts it, the design method, of a participant observer. In Glanville's own theory of objects, an object implicates an observer and vice versa. Surely conversation theory is bound up with participant observation, however formal (as it often is) and however impartial (as it may try to be). Projects in the OOC programme are suited, though not exclusively, to this approach and its methodologies.

What is the truth of a statement about a simple conversation ? It is an agreement or agreement to disagree, both of which are formalized as a dynamic form of Rescher's Coherence truth or equisignificantly by Taylor's interpretation of 'hermeneutic truth' as an iterative and continual 'refinement of meaning'. That, it may be added, is both the truth and the existence of a concept, question, or individual, as you prefer. It may be represented by a re-entrant form, generating a similarly re-entrant distinction (of concept, collection of them, individual, or whatever you should prefer). It is the essence of organizational closure and the greater closures obtained by the interaction of partially autonomous systems. At the level of P-Individuals, at any rate, it entails the gamut of issues raised by Bateson and Laing of 'your view of me' and of 'your view of my view of me', to cite but a couple of them. In the context of so-called Artificial Intelligence, there is a definitive paper by Gregory which aptly demarcates the mainstream of conventional machines. At very least it has demanded a different view of computation, or else something entirely novel.

These logical reframes are the offspring of different and multitudinous authors. But it is most commonly the forms of equations (or other correspondence relations) that are of particular interest (although numerical values assigned by some convention of spin, say, or orientation) are significant. In all cases, the lattice of truth values (see a paper by Jack Cowan, in the middle sixties) is richer than the Aristotelian form

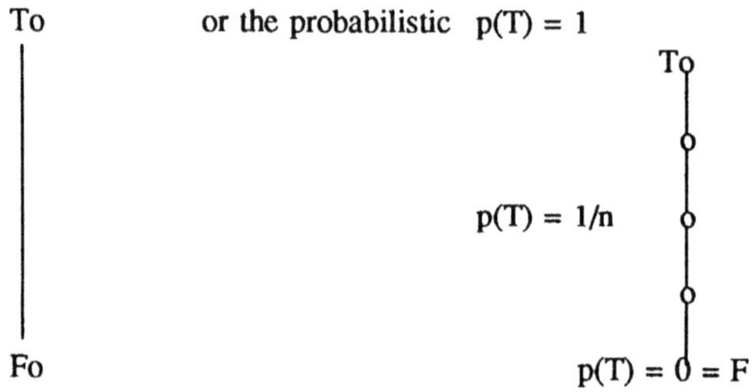

and the simplest are the Lewis possibilistic logics and the Deontic logics of Von Wright. Hellerstein recently proposed a logic, called 'Diamond' of superficial paradox. As with those already mentioned, 'Diamond' has a truth value lattice (that is, an arrangement of different *kinds* of truth valuation), which has the form

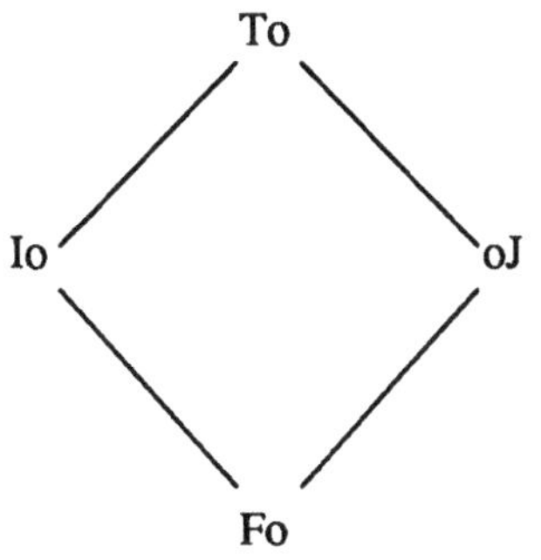

in one interpretation having I = <T,F,T,F,...>, J = <F,T,F,T,...> needed, for example, to characterize self-organization in the sense of Atlan (at the *new* symposium), and render it congruent with self-organization, according to Von Foerster. Another interpretation of 'I' and 'J' is 'self' and of 'other' referential, or the re-entrant values in the higher order calculus of distinctions, due to Spencer Brown et al. Greenberg makes similar proposals although (I think) he may need more kinds of truth functional. For certain, Louis Kauffman, in what I believe to be the most perspicuous system able to accommodate the *new* paradigm, based on an extension of knot theory to knots and twists with bifurcating and re-entrant strands, is in need of more. They can be supplied. For Günther, in some barely comprehensible but magnificently profound publications, has offered a system of logics which will surely provide at least a countable infinity of them, enough, for instance, to accommodate the values for metaphors designating analogies, the commands that are obeyed or disobeyed (not meta-linguistically terminated) and the questions which receive replies or maybe repartees (in Conversation theory, as formalized in Lp). It seems, as a reasonable conjecture, that a slight augmentation of the algebraic schemes that Kauffman has brought to the surface are able to give even that, most intractable of all fields, respectably manipulability.

4. Taxonomic indications

In view of the comments of the previous sections, I hope that these tabular comparisons and contrasts will make useful sense. They are terse and arranged in a somewhat arbitrary way. But they are an attempt to give a digestible indication of an evolutionary, rather than revolutionary, paradigm shift that has taken place, from the *old*, to *new*, as titled below:

OLD	*NEW*
INFORMATION Selective or Statistical	Coupling of organizationally closed distinct systems
CONTROL Of external variation	Increase or decrease distinction
COMMUNICATION Of message, data	Conversation. Concepts may be shared, personal concepts enriched
STABILITY Fixed points cycles, in state space of observer, Transition occurs, as in Ashby, meta-stability	Organizational closure. Boundary cycles, boundary toruses, in self or system determined system. Both classical attractors and chaotic attractors, the latter giving variety or distinction
RANDOMNESS Haphazard external or by basic inability of observer to discern	Randomness in previous sense, possibly, chaos essential, quite different. Full type of stochastic.
CAUSALITY, TEMPORALITY Newtonian Map, unique, as in system state trajectories. Transition through multiple causality, as in Singer	Causality, Temporality, as before but also in broader sense, for example, of Petri nets or Systems of Holt et al.
DISTINCTION By external observer, may be on good common-sense grounds	By closure, of systems. Is, for example, complementary to productive process. May be supplemented by observer
OBSERVATION Essentially External, Impartial, value free	Essentially Participant, entails values
CONSCIOUSNESS Exists but is excluded or considered in metatheory	Accommodates consciousness, for coming to know, personal or shared and re-entrant process

EPISTEMOLOGY Varied, but knowledge is usually seen as there, in data, facts, rules, and so on	Unless qualified, is act of coming to know, personal or shared and public concepts
LOGIC Typically Aristotelian or probabilistic type	Modal, of distinction, coherence, of process and re-entrant process

5. *Desiderata*

The *new* paradigm is less familiar. However, the *new* paradigm does satisfy certain desiderata for a well formed theory of wider scope.

The desiderata satisfied are reminiscent of a science like physics. Manifestly, however, they are *not* the same as those to be discovered in physics. The data are usually of a different type, hypotheses are differently formulated (notably by a participant observer) and tests of these hypotheses are supported or denied by evidence having a generally different truth valuation (coherence truth, for example, as in Rescher; hermeneutic truth in the sense of Taylor, iterative refinement of meaning, rather than the T, F, or probabilistic values of veridical truth). It can be held that coherence values are logically prior to the establishment of veridical or factual values, in any domain of interpretation, but in any case they are different.

Within the *new* paradigm, at least, we *do* encounter desirable features like very rigorous complementarities, such as process/product; exclusion principles; several conservation principles; for example, of information transfer, of activity and distinction; principles of parity and of duality. In brief, although broadening the compass of the paradigm of Cybernetics, it has been possible to increase its power, utility,its elegance, its beauty.

The *new* Cybernetics is vastly concerned with consciousness, life, the universe and all that which is submerged as private thought or unseeable, like the productive processes, beneath awareness, that are responsible for reproducing and maintaining the dynamics of a conceptual structure and its complementary architecture. It is a heady domain and one in which it is wise to be chary of mystics, as the Church is chary of canonizing her Saints. It is easy to parade the inconsequential as much vaunted without, most often, a shade of deceit, however, with the caveat of due and over-critical scrutiny, consciousness, life, the universe and all that are the business of this science. Although there is continuity between the original Cybernetics and the present form, the first Cybernetics disavowed consciousness as its concern, explicitly so in a paper by Ashby in a psychiatric journal.

Some, like the computing machine in the 'Hitch Hikers Guide to the Galaxy' say the answer to these questions is 42. Some say the answer is 63, but these are 'numerals, not numbers', so merely tokens.

From a participant, and likely enough biased point of view, I am fully convinced that these questions set for Cybernetics to ponder upon are crucial to the survival and the evolution of consciousness, life, love, the universe and all the rest; further, this is a demonstrable fact.

It is no longer necessary to prop up each novel move by arbitrary axioms. It is no longer warrantable, if ever it was so, to contend that Cybernetics might be no more than a useful bag of methods, open to scavenging whenever some tool is needed.

Appendix 1: The New Cybernetics and Chaos-theory

In the jargon of dynamic system theory, as witnessed by the recent coalescence of volumes devoted, exclusively, to this subject. Roughly speaking, if the behavioural trajectory of a dynamic system is plotted in a standard state space, or the multiple trajectories are plotted upon a phase space, they commonly reach 'attractors', or since dynamism is conserved, 'cycles' or more complex series of dynamic attractors. Of these, there may, of course, be many and, in deterministic cases, those approached will depend upon the condition of the system, at the outset, formally one state representative point or many of them.

A very elegant, but also readable account, is given by Pietgen et al. It also demonstrates (a) that not all systems, notably those characterized by the indefinite iteration of fundamentally non linear equations of motion, approach attractive cycles of different kinds. These cycles are known, in the academic and popular literature, as chaotic, chaos meaning the generation of variety from *within* a system, in contrast to random perturbations from *outside* the system, hence unpredictable, and indeterministic due to arbitrary perturbations. However, chaos, in this sense, is entirely deterministic, yet altogether unpredictable, a peculiar feature, one suppressed as pathological by the establishment, until quite recently, although it has been well known since, at least, Poincaré, and probably for many ages before that, (b) that, in order to deal with chaotic attractors, amongst which, as in the minimal case of the Lorentz attractors, there are many competing and/or cooperating attractors. Those in competition and/or cooperation, be they normal or chaotic, are truly stable, that is, dynamically equilibrial. After small and arbitrary perturbations of a system, the system returns to its dynamic equilibrial form. The most familiar scenario of approach to this condition, is the Fiegenbaum or Frequency doubling scenario, where trajectories cross each other but remain equilibrial.

To accommodate these curious, but entirely formal modes of dynamic system behaviour, it is possible to adopt various expedients. One is to deform the usually Euclidian descriptive space, state space or phase space, into rational although odd shapes. On deeper scrutiny, this expedient is entirely congruent with asserting a complementarity between a process, casually represented as a line-like trajectory, and its product, which is what may actually be described. Once again, and rather crudely, the process implies a product, boundary, carapace, or a distinction. It is, for example, the distinction between the *inside* and the *outside* of a system, that which makes it a system of necessity, although observers, of course are at liberty to choose any reasonable distinction. This one is, however, canonical. It is a distinction necessarily created by the processes it contains. If only because of that it is quite legitimate to allude to 'random' perturbation coming from outside the system, as an unpredictable and indeterministic influence and 'chaos', which is deterministic, although unpredictable coming from within the system and maintaining a delicate balance. It was presciently stated by Ashby, many years ago, that a regulator or system must possess at least as much variety of its own as the environment in which it exists. At any rate, whatever stance you choose to adopt, the boundary, the carapace, the deep rooted distinction whereby the system exists is dependent upon this delicate balance, of 'random' perturbation and 'chaos' as an internal variety generator. Otherwise, come to think of it, there would be no canonically identifiable systems. But, common sense shows us that they exist.

Any universe of discourse, interpretation, or what you wish, is replete with them and filled with them. An observer, however impartial, is such a thing and so is the joint system formed by observation. The last point renders the observer and the observed interdependent and gives them reality. Moreover, it affirms that no universe is an amorphous splodge. Both coherence is conserved, and distinction is conserved. Unity, a great quality, is not merely an unutterable uniformity.

Appendix 2: The New Cybernetics and Connectionism

It has been cogently argued that 'connectionism', as typified by such artifacts as the 'connection machines', is a significant breakthrough and possibly yet another phase in the evolution of Cybernetics. Whilst I see the sense in the argument and the value of these reinvented artifacts (they were prevalent in the 1950's and early 1960's, as manifest by a plethora of publications and tangible, if large and slow devices), they seem (even today) to suffer from similar defects to those that afflicted the earlier, slower, and larger generation.

In the first place, there is a great advantage to be gained from the connectionist architecture and technology. There is also a most lamentable tendency to reduce this advantage to a mere triviality (a faster way of doing things, like pattern recognition, which are well within the compass of standard computing machines; all of this can be done, of course, but so may the much more which is sadly omitted). If connection machines demonstrate anything essential, then they must not be treated as faster kinds of standard computers. Their other potentialities must be recognized and used.

To my mind, at any rate, the issue is how they are used. Can they be used to realize rather than simply simulate organizationally closed but also informationally open systems ? Yes, I believe so. If so, then use them for this. Can they implement, for real, genuinely modal logics, especially of distinction, coherence and action, with truth values related to Günther logics and proof or demonstration methods derived from an embeddment of Kauffman knot theory, related to one or other direction of the unit unknot, with many surgical operations and singularities ? Yes, I believe they can be, and also for many other interesting things. If so, then these machines and, to some extent the philosophy of their architecture, constitute a breakthrough.

It may be big enough to mark a novel departure point in the evolution of Cybernetics. I am not sure and leave it as a sensible but open question.

* This introduction was first published, though in a reduced form, in *Communication & Cognition*, vol. 23, nos. 2-3: *The Old and New in Cybernetic Fashions.*

References

Ashby, W.R., 1956, *Introduction to Cybernetics*, Chapman and Hall, London.

Bar-Hillel, Y., 1964, *Language and Information Processing*, Addison-Wesley, Reading Massachusetts.

Bateson, G., 1972, *Steps Towards an Ecology of Mind*, Paladin, New York.

Beer, S., 1972, *The Brain of the Firm*, Wiley, New York.

Braten, S., 1980, Dialogic, *Proc. 3rd International ARI Symposium on Complex Decision Making*, System Research, Richmond, Surrey, ARI, Washington D.C.

Cherry, C., 1957, *On Human Communication*, Wiley, New York.

Couffignal, L., 1961, La cybernétique comme méthodologie, *Les Etudes Philosophiques*, vol. 16, no. 2, pp. 157-165.

Cowan, J.D., 1962, Many-valued logics and reliable automata, in *Int. Tracts in Computer Science and Technology and their Application*, vol. 9: Principles of Self-organization, H. Von Foerster and G.W. Zopf (eds.), Pergamon Press, Oxford, pp. 135-179.

De Zeeuw, G., 1986, Social Change and the Design of Enquiry, in *Sociocybernetic Paradoxes*, F. Geyer and J. Van der Zouwen (eds.), Sage, London.

Flores, F.; Winograd, T., 1986, *Understanding Computers and Cognition*, Ablex Publishing Company, Norwood, New Jersey.

Gabor, D., 1961, A Universal Non-Linear Predictor and Simulator which Optimises itself by a Learning Process, *Proc. Inst. Elect. Engineers*, vol. 10.

Glanville, R., 1976, What is memory that it can remember what it is, in *Progress in Cybernetics and Systems Research*, R. Trappl and G. Pask (eds.), Halstead Press, New York, pp. 27-37.

Glanville, R., 1980, The models dimensions, *International Journal Man-Machine Studies*, vol. 13, pp. 305-327.

Glanville, R., 1988, *Objekte*, Merve Verlag, Berlin.

Gregory, D., 1986, Philosophy and Practice of Knowledge Representation, in *Human Productivity Enhancement*, J. Zeidner (ed.), Praeger.

Günther, G., 1962, Cybernetic Ontology and Transjunctional Operations, in *Self-organizing systems*, M. Yovits, G. Jacobi and G. Goldstein (eds.), Spartan Books, Washington, pp. 313-393.

Hellerstein, N.S., 1985, Diamond, a logic of paradox, *Cybernetics*, vol. 1, no. 1, pp. 101-115.

Kauffman, L., 1986, Self-reference and Recursive Forms, *Journal of Social-Biological Structures*, no. 10, pp. 53-67.

Kauffman, L., 1987, *On Knots*, Princeton University Press.

Laing, R., 1961, *Self and Others*, Tavistock, London.

Maturana, H., 1978, The Biology of Language and the Epistemology of Reality, in *The Psychology and Biology of Language and Thought*, G. Miller and E. Lenneberg (eds.), Academic Press, New York.

McCulloch, W.S., 1965, *Embodiments of Mind*, The MIT Press, Cambridge, Massachusetts.

McKay, D.M., 1956, Epistemological Problems for Automata, in *Automata Studies*, no. 34, C.E. Shannon and J. McCarthy (eds.), Princeton University Press.

Pask, G., 1975, *Conversation, Cognition and Learning*, Elsevier, Amsterdam.

Pask, G., 1975, *Conversation Theory, Applications in Epistemology and Education*, Elsevier, Amsterdam.

Pask, G., 1978, Minds and Media in Education and Entertainment, in *Progress in Cybernetics and Systems Research*, vol. 4, G. Pask and R. Trappl (eds.), Hemisphere, Washington, pp. 38-50.

Pask, G., 1980, Developments in Conversation Theory, *International Journal Man-Machine Studies*, vol. 13, pp. 357-411.

Pask, G., 1982, Some Generalisations of Conversation Theory and the Protolanguage Lp, in *Progress in Cybernetics and Systems Research*, vol. 9, R. Trappl, L. Ricciardi and G. Pask (eds.), Hemisphere, Washington, pp. 407-424.

Pask, G., Review of Conversation Theory and the Protolanguage Lp, *Educ. Comm. Tech. Journal*, vol. 32, no. 1, pp. 3-40.

Pask, G., 1990, Aspects of Conversation Theory and Lp, in *Self-Steering and Cognition in Complex Systems*, F. Heylighen, E. Rosseel and F. Demeyere (eds.), Gordon and Breach, New York, pp. 240-247.

Petri, C.A., 1964, Communication with Automata, Translation F. Green, MIT, A Suppl. to the Techn. Rep. 1, Rome Air Development Centre, Contract AF.30 (604)3324.

Petri, C.A., 1972, Comments by Anatol Holt, in *Our Own Metaphor*, M.C. Bateson (ed.), A. Knopf, New York.

Pietgen, F.H.; Richter, P.H., 1982, *The Beauty of Fractals*, Springer Verlag, Bremen.

Rescher, N., 1973, *The Coherence Theory of Truth*, Oxford University Press, London.

Shannon, C.E.; Weaver, W.E., 1949, *The Mathematical Theory of Communication*, University of Illinois Press.

Spencer Brown, G., 1969, *The Laws of Form*, Allen and Unwin, London.

Taylor, C., 1964, *The explanation of behaviour*, Routledge and Kegan Paul, London

Taylor, C., 1973, Interpretation and the sciences of Man, *Review of Metaphysics*, vol. 25, pp. 3-51.

Thom, R., 1975, *Structural Stability and Morphogenesis*, Benjamin, Reading, Massachusetts.

Varela, F.J., 1979, *Principles of Biological Autonomy*, North Holland, New York.

Von Foerster, H., 1960, On self-organizing systems and their environments, in *International Tracts in Computer Science and Technology and Their Application*, vol. 2: Self-organizing Systems, M.C. Yovits and S. Cameron (eds.), pp. 31-51.

Von Foerster, H., 1981, *Observing Systems*, InterSystems Publications, The Systems Inquiry Series, California.

Von Wright, G.H., 1963, *Norm and Action*, The MIT Press, Cambridge, Massachusetts.

Wiener, N., 1948, *Cybernetics*, The MIT Press, Massachusetts.

SELF-ORGANIZATION AND COMPLEXITY

ENDS AND MEANING IN MACHINE-LIKE SYSTEMS*

Henri Atlan

Hebrew University Hadassah Medical Centre
Kiryat Hadassah, Jerusalem
Centre Hospitalier Universitaire Broussais-Hotel-Dieu
Paris

Abstract

The meaning of instructions in a computer programme or of relations between parts in a functioning system is usually defined in relation to the task to be performed. However, contrary to man-made artifacts, the problem of the origin of meaning in natural systems must be approached without assuming end-seeking processes where explicit final states must be defined a priori.

In order to make the notion of meaning more precise, the theory of algorithm complexity is used to define a quantity called 'sophistication'. A clear distinction between programme and data is used for this purpose. Sophistication as a minimum programme length is a measure of the meaningful complexity of a machine assumed to be defined by its function itself, viewed as the outcome of a computing algorithm. The metaphor of the genetic programme presumably embedded in DNA strings is discussed in this context. It is opposed to the alternative hypothesis of DNA as data processed by a programme embedded in the whole cellular biochemical machinery.

The emergence of meaningful functions in the form of classification procedures is shown to happen in functional self-organizing automata networks. This suggests the possibility of mechanisms whereby meaning can emerge in natural systems without having to define a priori the ends the systems are seeking. In such instances, the order between end and meaning seems to be reversed: meaning is created first as a classification criterion and the function is defined after or at least simultaneously in relation to it, as the classification according to that criterion.

Machine-like systems are either actual machines built by men to perform given tasks, or natural organized systems which appear to perform some functions and, we assume, can be understood on the basis of the principle of causality

G. Van de Vijver (ed.), New Perspectives on Cybernetics, 35–48.

without, however, resorting to any kind of free will. In both cases the meaning of the organization — i.e. of the spatio-temporal relations between the parts — is derived from the task to be performed at the end.

However this task is known and explicitly defined in man-made machines while it is ignored in natural organizations, where its existence as a directive principle in the building of the machine is questionable. Nevertheless, until recently one man-made machine, namely the general purpose computer, has been used, with a few exceptions, as the only thinkable model to understand the organization of complex machine-like systems. This was felt particularly in biology and in psychology.

In biology the computer was the basis (and still is for most biologists) of a very powerful and dominant metaphor: the metaphor of the genetic programme. According to this representation in its extreme form, every individual is determined by a computer-like programme, written in the DNA sequences of its genes. Its development, its growth, its reactions to the environment acting as a source of stimuli, are the results of the execution of this programme. This programme has been written during the evolutionary process by natural selection, so that the organism is adapted to its environment; i.e. it reacts to its environment by subroutines written in its genetic programme, and is called by given stimuli.

Of course, already from the beginnings of molecular biology — and this can be found in every textbook — we were warned not to take the metaphor too literally: the genetic programme was described as a very peculiar programme since it needs the products of its reading and execution to be read and executed; or else, it functions as a self-programming programme. Additional questions were raised about the language in which it is written and it was recognized very soon that the genetic code certainly does not constitute a computer language, being at most a lexicon.[1]

Nevertheless, the computer model has been useful and is still useful for its heuristic value, giving a framework to describe in a convenient way known and as yet unknown genetic determinations.

However, the computer model has a big disadvantage: at the end, one finds in it the very thing one wanted to eliminate at the beginning, i.e. one or other difficult aspect or property in the crude, observed phenomena which was difficult to account for by a causal explanation, so that the model was built in order to eliminate it. I am thinking for example about purposefulness in biology and subjective introspective data in psychology, that the computer models are supposed to eliminate.

In biology one wanted to eliminate the apparent purposes in the observed development and adaptive phenomena by sticking to mechanistic processes, but one does find them in the computer model since any computer programme has always been written with a given specific purpose in mind.

In cognitive sciences also it seems that one wants to eliminate the subjective introspective nature of human experience but one finds it disguised within the semantic components of any computer-like model, where the semantic content must be fed from the spontaneous natural human experience.

This is due to the fact that the computer is not a natural system, just given by nature to be used as a model for other natural systems like the living organism or the human cognitive system. As a matter of fact, it is only a particular instance, a particular kind of production of human cognitive systems and it should not be more legitimate to use it as a paradigm for how the human cognitive system functions than any other production of a human cognitive system, such as a metaphysical theory, or a philosophical theory, or any system of interpretations that man can imagine and use to put some order into his experience of the world.

It is certainly an interesting problem in the history of science to understand why the computer model has nevertheless imposed itself not only in human sciences but also in basic biology, after a century during which the central model was the steam machine, which itself had replaced the watch and the pendulum. As I said before, the metaphor of a computer programme in basic biology was meant to account for apparently finalistic genetic determinations. The word teleonomy was invented in the fifties by Pittendrigh to replace teleology, in order to get rid of finalism in biology.[2] Apparently end-seeking processes are observed in adaptation and development and classical biology was always mixed up with teleological, i.e. finalist thinking. In order to get rid of it, the goal of research was stated as looking for mechanisms with teleonomy, the difference being that in one case (teleology) one deals with purposeful and end-directed processes governed by an 'intelligent, designing mind' while in teleonomy one is looking for non-purposeful end-directed processes. The computer programme came at precisely the right time to provide a model for such a non-purposeful end-directed process: the computer executes a programme, mechanistically, without purpose, although it works as an end-directed process. The only problem, of course, is that in the metaphor of the genetic programme the existence of a programmer and his purpose are forgotten; of course the nature of the programmer who wrote the genetic programme is left outside of the scope of the metaphor ... unless one wishes to state that natural selection works as a programmer writing programmes. Then one is led back to the purpose always present in the work of a programmer.

The reason that theories of self-organization were sought for was to try to avoid these difficulties. The question was: which logical prerequisites must be met for a self-organizing system to exist ? I.e. a system where no programme would have to be fed from the outside by a programmer, or a system working like a self-programming programme ?

So far, self-organization theories have been concerned mostly with the emergence of structure, the creation of complex systems from less complex ones, and very little with the emergence of function and meaning. The latter problem is certainly the most difficult one. It is related to a fundamental question which can be stated in different ways: how can a goal emerge as a source of meaning in a self-organizing process ? Or else, what about intentionality in machines ? Intentionality is experienced within our human psychological environments and we tend to project it into animals, at least into some of them like monkeys, dogs, horses, and so on, not so much into others such as amoebae, flies, which we try to analyze only in mechanistic terms without resorting to intention.

However, in an attempt to bridge the gap, at least theoretically, between these two kinds of experience, the question is raised whether *intentionality* is possible in machines ? As you know, Searle's answer to this question is definitely yes by definition, because we, including our minds, are assumed to be some kinds of machines, and we experience among ourselves intentionality as "a causal source of action".[3]

However, it is clear that we do not experience intentionality in all kinds of machines, at least not in man-made machines and it would be difficult without resorting to *animist* metaphysics to assign intentionality to all kinds of organizations, such as a star or a liquid vertex or a living cell.

Thus we are left with the question: which kind of machines can exhibit intentionality ? So far, it seems that the only answer we can give, in a circular way, would be: "those particular kinds of machines endowed with biochemical organization where a brain like ours has evolved".

As I said, this kind of statement seems to be circular and not very informative since we do not know how our brains generate, so to speak, intentionality.

However, we can make a small step forwards when we realize, as many of us here have stressed already, that a crucial feature of biochemical organization is its hierarchical nature. There are several levels of organization; and most of what we have no knowledge of at present in the field of biology has to do with the articulations between one level and another.

Therefore our question can now be formulated in a different way.

Since a network of interacting automata is a simple but good enough simulation of a hierarchical organization where the articulations between local and global properties can be studied, the question is now: how can intentionality emerge in a self-organizing process as a global property of a network of interacting automata ? There are some obvious necessary conditions which self-organizing processes have in common; among them, the building of the network must include some indeterminacy, some randomness in one way or another, either in the rules themselves or in setting up the initial conditions, or in the connections, i.e. some noise must be active at some level, in one way or another.

But what else ? Before trying to answer the question let us consider the following.

Let's assume for a moment that we can answer the question completely, i.e. that we can construct or simulate a self-organizing system which would behave in an intentional way as 'a causal source of action'. Then, we will be left with another question. Is this kind of intentionality which emerges from self-organizing networks of automata the same as that which we experience in our human environments ? I think that this last question will always be undecidable because it will never be possible to make our *subjective* introspective experience of intentionality and decision making coincide with our *objective* observation of the behaviour of a self-organizing machine. It will always be possible to decide, arbitrarily in some sense, that intentionality similar to ours exists only in those particular machines with which we are connected in a network of relationships involving moral or at least social and legal *responsibility*.

Therefore, coming back to our question within this limited scope, it can be reduced to a less ambitious one, namely "How can automata networks exhibit self-organizing properties in the *strong sense* of creating not only the process, the path to be followed in order to achieve a given goal, but to create the goal itself ?".

This amounts to the self-creation of function and meaning in organization, not only of structure.

Although we still don't know exactly what we mean in general by meaning, we can restrict ourselves and try to formalize the generation of meaning which has to do with *performances of functions*. I am referring not only to physiological functions of organisms but also to performances of machines and algorithms where the meaning of a part or of a programme instruction has to do with its function, i.e. with what it is good for within the overall functional organization — the flow chart if you wish — of the system.

However, again, it is important to remember that the meaning of information in natural organizations such as biological systems is very different from what it is in artificial constructs. In man-made machines or programmes the goal is defined from the beginning and it is in relation to this goal that parts, connections, signals, etc., either have a meaning or are meaningless. In a natural organization, the final goal, if there is one, is not known, and meanings are defined a posteriori according to *self-generated* criteria. This is why a real understanding of self-organization amounts to understanding mechanisms by which *non-purposeful systems*, not goal-oriented from the outside, organize themselves in such a way that the meaning of information is an *emerging* property of the dynamics of the system.

Before attempting to show you how simulations of automata networks can help to visualize such mechanisms, I should like to spend some time on a recent attempt, in collaboration with M. Koppel, to formalize meaningful complexity, keeping within the framework of the theory of algorithms.[4]

The approach is based on the following question: faced with a natural phenomenon which seems complex to us — because it does something or for some other reason — is there a way to decide to what extent we are dealing with randomness or with a complex meaningful structure ? In other words, in what we observe, what part is played by randomness and what part by meaningful structure ?

This question is not only a formal one: for example if we receive signals from outer space or more practically when we look at DNA sequences, is there a way of knowing how much there is of meaningful complexity that is worth investigating ?

What is the relationship between the intrinsic properties of a string — e.g. the outer space signals or the DNA sequence — and its function, i.e. its effect on a receiver, that usually tells us about its meaning ? In our preliminary attempts to answer this question we based ourselves on a modified version of the classical Kolmogorov-Chaitin theory of algorithmic complexity.

According to this theory, the complexity of a string — such as either of the ones I just mentioned — is the length of the minimal description capable of reproducing it when it is given to a normalized general computer, i.e. a Turing machine. This description is itself a string which can be written in a binary alphabet, and the number of bits is a convenient measure of its length.

Now, in this description, no distinction is made between what is a programme and what are data.

When we are dealing with usual man-made procedures, the distinction is obvious: a set of operations is a programme and the data on which it operates are the data.

However, for the theory, it does not make any difference, because for the machine itself it does not make any difference: the bits of a programme are themselves data for the interpreter of the machine and therefore it seems that there is no intrinsic difference between computer programmes and data. In fact, the realization that there is no intrinsic difference for the machine between the programme and data was a great achievement which has led to highly significant breakthroughs in the theory of computation.

At the same time, however, something essential has been lost. Because this is true, again, only if we agree to put aside the distinction between what is meaningful and what is not in the minimal description of a string. If we wish to capture the meaningful structure of the string, we must come back to the distinction between programme and data, and make it even more clear cut. What is at the root of this distinction is the fact that a programme has a certain *range*, i.e. the class of all the objects or strings it can produce when it is given all possible data. What is *common* to *this class* of objects is their structure or meaningful complexity, whereas each particular element of the class is a particular instance of this structure.

Thus, instead of considering a usual Turing machine, we consider one with two distinct input tapes — one for the programme and one for the data. And the minimal description of a string is thus made up of two parts, programme and data, each with its length, so that the usual complexity measure H(S) becomes the minimum of (length of programme P + length of data D), (P,D) being a description of the object S:

$$H(S) = \min \{ |P| + |D| \mid (P,D) \text{ generates S when input to a universal Turing machine} \}$$

Now, what we call the meaningful complexity of S, or, as we called it, its sophistication, is only the length of P, i.e. the length of the programme part in the minimal description. Thus, we succeed in sorting out the meaningful complexity in such a way that a random string, where the programme part is zero, has a zero sophistication, while its classical complexity is high.

In fact it is not that simple because the fact that data and programme together can be regarded as one single input, i.e. data, to a universal Turing machine, has deep justifications from the computational point of view, which have to do with the role of the Turing machine as an interpreter.

Therefore, it may be that a minimal description is minimal only for a given particular string and not necessarily for the whole class of strings which share the same structure; in other words it is possible that this particular minimal description is *ad hoc*; and if we want one which is not, and which really captures the structure of the class, we may want to allow the description to be longer but appropriate for the whole class.

This is why we introduced the idea of c-minimal descriptions, made up of (P,D) so that:

$|P| + |D| \leq H(S) + c$, i.e. we allow for the description to be longer than the minimal by a constant c.

And the sophistication or meaningful complexity is defined in relation to this constant as c-sophistication:

$SOPH_c(S) = \min \{ |P| \mid \exists D,\ |P| + |D| \leq H(S) + c\}$, i.e. the minimum programme length P so that (P,D) is a c-minimal description of S.

The role of constant c is better understood by an example. The following are three possible different descriptions able to generate the characteristic string of prime numbers.

1.*(PRINT,DATA) =*
PRINTS 1,2,3,5,7,11,13,17,...
(NEEDS DATA ON PRIMENESS OF ALL NUMBERS)

2.*(P2,ODD DATA) =*
FIRST TESTS FOR EVENNESS TO ELIMINATE EVEN NUMBERS (GREATER THAN 2).
THEN PRINTS THE PRIMES FROM THE DATA ON ODD NUMBERS
(NEEDS DATA ONLY ON PRIMENESS OF ODD NUMBERS)

3.*("PRIME") =*
APPLIES ON EVERY NUMBER A TEST FOR PRIMENESS
(DOES NOT NEED ANY DATA)

Let us try to find the sophistication of that string by looking at what its minimal description is. This obviously will depend on the length of the string we are interested in. For a short string of a few prime numbers, the first description, namely PRINT + the DATA (i.e. the string itself) is certainly the shortest description — shorter than the programme *PRIME*.

However, the longer the string, the longer the data necessary for that description (i.e. the whole string), and the more economical the actual programme *PRIME* appears, because it requires less data, as a matter of fact, none at all.

Intermediate between them is P_2: the additional programme length to test for evenness is economical when the length n of the string becomes longer, so that the n/2 bits of data which are saved by the test compensated for this additional length.

Therefore, we do not always want the true minimal description such as PRINT + FEW DATA which might be ad hoc and appropriate only for a given set of data corresponding to only a few members in the class of strings which share the same structure. We allow the description to be c-minimal, i.e. longer than the minimal by a constant c. This allows for additional length of programme which will be confirmed when additional data are given. In that sense, c is a measure of confirmation or redundancy. If a programme is the true one, it will remain the shortest even for large values of c, i.e. for long strings.

Thus, the constant c measures what we are ready to accept as additional bits in order to be confident that we are not dealing with an ad hoc minimal programme, i.e. a programme which works only for a given set of data.

One of the advantages of this definition is that it can be extended to infinite strings; and then it can be shown that the parameter c is eliminated.

Thus, a finite sophistication can be defined for an infinite string. It can also be shown that its value is invariant upon the choice of the universal Turing machine.

In cases where the length of the minimal *programme* increases and tends to infinity, when the length of S tends to infinity we are dealing with an infinite sophistication and this defines in a precise way an object which is *neither* recursive nor random. It is worth noting that this strange kind of object might be the kind of thing nature seems to manufacture normally.

These definitions have been of some help in analyzing the question of what is the computing nature of the DNA strings: are they programme or data ? According to the widespread metaphor of the genetic programme, DNA are viewed as programmes computed by the cell. When the metaphor is taken too seriously, it leads to the false assumption that knowing the DNA sequence of a species is enough to describe its structure and function. It seems more likely that DNA strings should be looked at as data fed to the cell metabolism machinery which would work like a programme. If that is the case it is obvious that the main theoretical assumption which underlies the much publicized project of Sequencing Human Genome is wrong.

In conclusion, I would like to go back to the emergence of meaning in functional self-organization to show you that it is not so mysterious, since it can be simulated relatively easily by networks of automata.

This work was done in collaboration with G. Weisbuch and F. Fogelman, and made use of Boolean networks which were first studied by S. Kauffman.[5]

In a typical situation, networks made of connected Boolean automata are constructed as follows: each automaton is located at a node of a 16 x 16 matrix closed on a torus; it receives two input connections from two of its opposite neighbors and sends two output connections to its two other neighbors.

Each automaton is characterized by a given Boolean function of two variables. At every step each automaton updates its state by computing its function from its two binary inputs and its new state is sent as output. The 14 non-constant two-variable Boolean functions are distributed randomly on the nodes of a matrix. Initial states are also set randomly and the automata compute their new states in parallel, at every time unit.

Well-established results have shown that in most cases the network reaches a limit cycle relatively rapidly. In its limit cycle the network presents a macroscopic spatio-temporal structure since it is divided into connected subnets, stable and oscillating. An example is shown in Fig. 1, where S represents the stable elements and P the periodically oscillating elements.

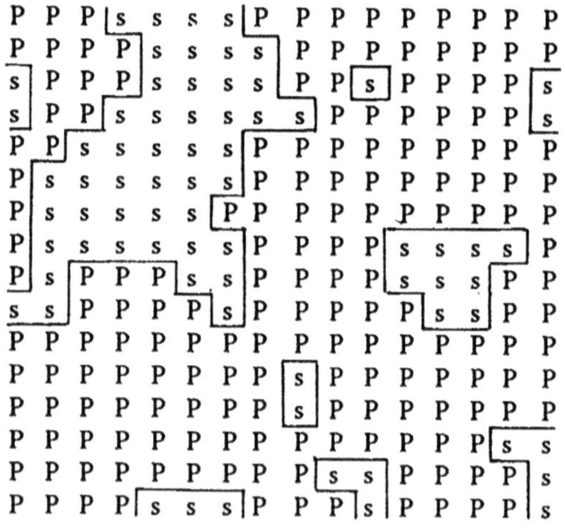

Figure 1

This process is an instance of structural self-organization with the emergence of macroscopic structures from local interactions, which has been extensively studied under different conditions as far as the nature of the automata and of their connections is concerned. Less known is the utilization of such networks for simulations of functional self-organization. When such a net is in a limit cycle, temporal binary sequences are imposed on one automaton and work as 'permanent' perturbations of the limit cycle. By 'permanent' we mean a duration of two cycle lengths, plus an additional two cycle lengths, during which the perturbed state of the networks is observed. Because of its robustness the macroscopic structure of the limit cycle is very often grossly conserved (depending on the location of the perturbed automaton), with the exception of a few stable elements which are destabilized and also, surprisingly enough, some oscillating elements which are stabilized by the perturbations. Since stabilization of a given oscillating element by imposition of perturbating sequences on another one occurs only for a given class of such sequences, this process can be used as a sequence recognizer. The interesting feature in this phenomenon is the definition of the criterion for the classification of sequences into those which are recognized (by the fact that they stabilize a given element) and those which are not. An example of such criterion is shown in Fig. 2. Classes of binary sequences are defined by partially random periodical structures where indifferent digits (as those indicated by stars in Fig. 2) are inserted at given locations. The definition of such a structure is the result of a given pathway in the network between the input element where the perturbating sequence is imposed and the output element which is (or is not) stabilized by the sequence.

a	00101000100000011010000110001001
b	10001001000010010010100100101000
c	01001100100010001011010001000101
d	10100001101010010010100110101000

* 0 * 0 * 0 0 * * 0 * 0 * 0 0 * * 0 * 0 * 0 0 * * 0 * 0 * 0 0 *

Figure 2

Out of the four strings represented, only a,b,d, belong to the class of pseudo-periodic, partially random strings defined by the criterion *0*0**0*

The existence of 'star' indifferent elements is the result of the overall spatio-temporal structure of the net in its limit cycle and of the 'reducing' effects of computation by some of the Boolean functions which associate one identical output to three different two-variable inputs, as is shown in Figure 3.

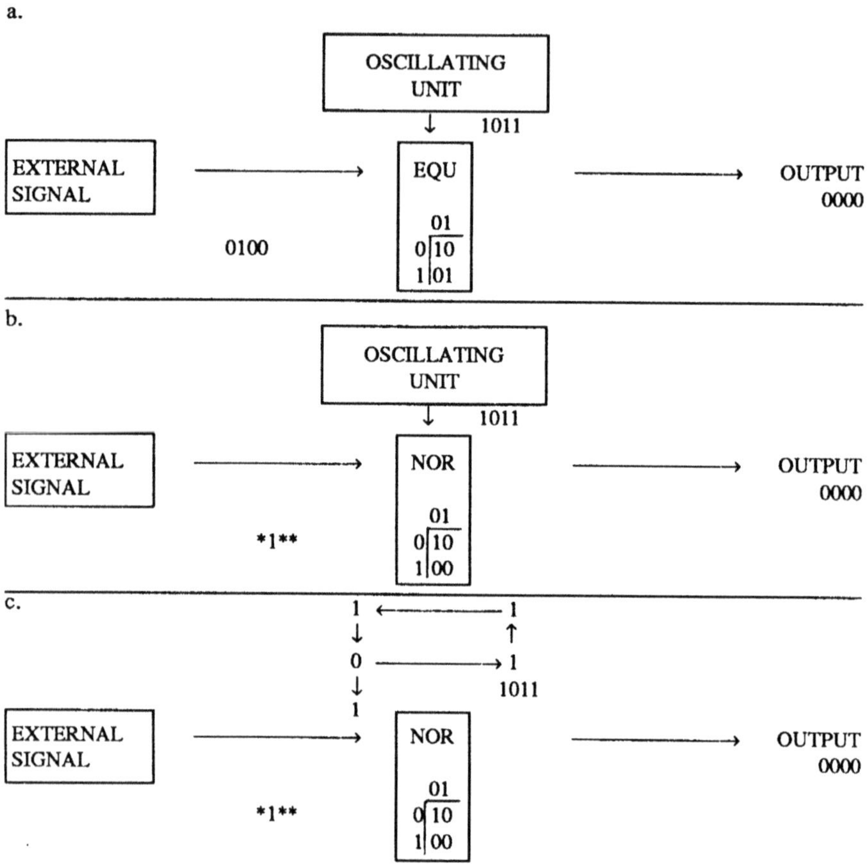

Figure 3

Local mechanisms of stabilization.[6]

a. A previously oscillating EQU element can be stabilized into a constant 0 state by a perturbating sequence applied to one of its inputs which matches exactly the sequence on the other input.

b. This NOR element recognizes a class of partially random sequences. The first, third and fourth bits of sequences can have any value, since the input from the oscillating unit is 1 and the output therefore will be 0 in any case. Only the second bit has to be 1 for the output to be still 0.

c. A recognition device can be built from a generating frustrated loop made of transfer functions (2-to-1 reducing) associated with a 3-to-1 reducing function for fuzzy recognition (classes of partially random strings), or a non-forcible function (EQU, XOR) for precise recognition (truly periodic strings).

Thus, a specific pathway in the network, after it has reached its final structured state, between a given element serving as input and another one serving as output, defines a class of sequences to be recognized. In other words, the structural self-organizing process results in a non-programmed definition of criteria for classification of binary sequences.[7]

This model shows a mechanism by which a set of messages is divided into those that are recognized and those that are not, while the criterion for this demarcation — which is similar to making sense and not making sense to a cognitive system — is nothing other than a given inner structure, which has no other meaning than being able to produce this demarcation. The criterion itself may have come about, at least partly, in a random manner.

It is as if complexity, which appears as an apparent non-reducible randomness, can be removed by means of a kind of orderliness that did not come about as a result of planning but as a result itself of apparent indeterminacy and randomness. This, in my opinion, is the consequence of the close relationship between complexity and disorder in natural systems not planned and ordered by man, the only difference being the existence of an apparent meaning or function in the former to the eyes of the observer.

This may also explain the feeling that we have about how nature seems to build its machines. This has been described by François Jacob as 'bricolage' ('tinkering'). To perform a complicated task, such as embryonic development or evolution of species, nature does not seem to work the way we would have done it if we had had to plan for it. It seems to take what happens to be at hand and to do something with it depending upon the circumstances, sometimes in a very cumbersome, and not necessarily the most economical, manner.

We get the same feeling when we look at our networks and see the kind of criteria which happen to be used in order to distinguish between classes. They also seem to be far-fetched from our point of view, if we would have had to set up a priori such criteria for what makes sense and what is meaningless.

* This article was first published in *Communication & Cognition*, vol. 23, nos. 2-3.

Notes

1. B. Shanon, 1978, The Genetic Code and Human Language, *Synthese*, vol. 39, pp.

401-415. See also: D.R. Hofstadter, 1979, *Gödel, Escher, Bach*, Basic Books, N.Y.

2. C.S. Pittendrigh, 1958, Adaptation, Natural Selection and Behavior, in *Behavior and Evolution*, A. Roe and G.G. Simpson (eds.), Yale University Press, New Haven, Connecticut, pp. 390-416.

3. J.R. Searle, 1980, Minds, brains and programs, *The Behavioral and Brain Sciences*, vol. 3, no. 3, pp. 417-457.

4. M. Koppel, 1987, Structure, in *The Universal Turing Machine, A Half-Century Survey*, R. Herken (ed.), Oxford University Press, London; H. Atlan and M. Koppel, 1990, The Cellular Computer DNA: Program or Data, *Bulletin of Mathematical Biology*, vol. 52, no. 3, pp. 335-348; M. Koppel and H. Atlan, 1991, Les gènes: programme ou données ? Le rôle de la signification dans les mesures de complexité, in *Théories de la complexité*, Colloque de Cerisy, Seuil, Paris, pp. 188-204.

5. H. Atlan, F. Fogelman-Soulié, J. Salomon, G. Weisbuch, 1981, Random Boolean Networks, *Cybernetics and Systems*, vol. 12, pp. 103-121; F. Fogelman-Soulié, E. Goles-Chace, G. Weisbuch, 1982, Specific roles of the different Boolean mappings in random networks, *Bulletin of Mathematical Biology*, vol. 44, no. 5, pp. 715-730; F. Fogelman-Soulié, 1984, Frustration and Stability in Random Boolean Networks, *Discrete Applied Mathematics*, vol. 9, pp. 139-156; H. Atlan, E. Ben Ezra, F. Fogelman-Soulié, D. Pellegrin and G. Weisbuch, 1986, Emergence of Classification Procedures in Automata Networks as a Model for Functional Self-Organization, *Journal of Theoretical Biology*, no. 120, pp. 371-380; S. Kauffman, 1969, Metabolic Stability and Epigenesis, in Randomly Constructed Genetic Nets, *Journal of Theoretical Biology*, no. 22, pp. 437-467.

6. From H. Atlan, E. Ben-Ezra, F. Fogelman-Soulié, D. Pellegrin and G. Weisbuch, Emergence of Classification Procedures in Automata Networks as a Model for Functional Self Organization, *Journal of Theoretical Biology*, no. 120, pp. 371-380, 1986.

7. H. Atlan, 1987, Self Creation of Meaning, *Physica Scripta*, vol. 36, pp. 563-576.

HIERARCHICAL NON-EQUILIBRIUM SELF-ORGANIZATION AS THE NEW POST-CYBERNETIC PERSPECTIVE*

Stanley S. Salthe

Natural Systems
New York

Abstract

Natural systems are not well described by cybernetics, which is an equilibrium perspective on unchanging systems that are open only to knowable or stereotypical aspects of their environments. Natural systems are all dissipative forms, driven into existence by energy flows at the boundaries of other systems higher than them in scale. Once they succeed in emerging, they undergo a stereotypical *development* from immaturity to maturity, and are then recycled. During this time they traverse a series of stages described as a *specification hierarchy*. Open systems like this also *evolve* in the sense of being irreversibly marked by traces from encounters with perturbations from their environments. Cognition in such developing-evolving systems obviously cannot be well described cybernetically. The purpose of this paper will be to criticize existing cognitive simulations as being non-verisimilitudinous to natural systems.

Natural systems are not well described by cybernetics, which is primarily an equilibrium perspective on systems that are open only to knowable or stereotypical aspects of their environments (so-called inputs). The implicit (scalar) hierarchical content of such systems has been explicitly explored in my *Evolving Hierarchical Systems* (Salthe, 1985). Causality in such systems is much richer than one might guess from most cybernetic matrices, and may be approached by way of extensions of the Aristotelian causal dichotomies, material/formal and efficient/final (Figure 1). Such systems are not well described with less than three *levels of organization* (the focal level, the lower level that gives rise to it and from which it *emerges*, and the contextualizing higher level, which regulates, calls for, or interprets it). It has been suggested by Ulanowicz (*Syst. Res.*, vol. 6) that in self-organizing systems, formal causality is found as well at the focal level itself.

G. Van de Vijver (ed.), New Perspectives on Cybernetics, 49–58.

THE BASIC TRIADIC SYSTEM

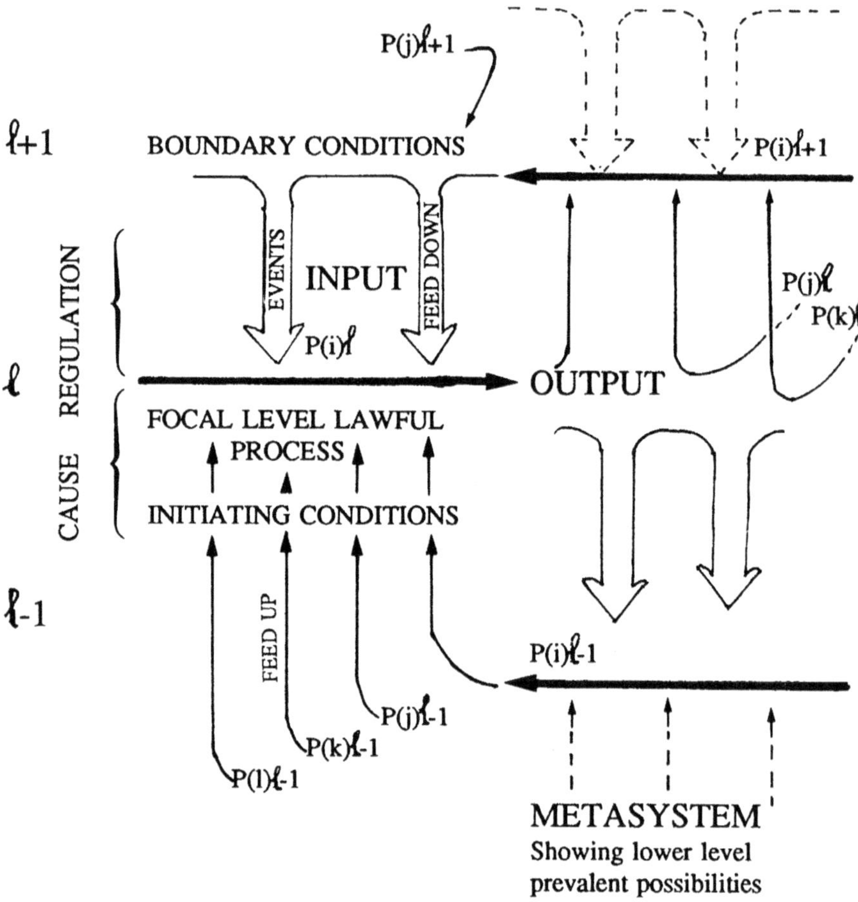

Figure 1

Visual representation of abstract relationships (no holons shown) in a scalar hierarchy

Let us examine Boolean networks (as in Atlan et al., 1986) from this point of view (Figure 2). Lower level constraints (initiating conditions of Figure 1) are represented by the Boolean functions themselves. Higher level constraints (boundary, including initial, conditions) are represented by the lattice of relationships between automata, and by any perturbing inputs that may affect the system. (One might note that still higher level constraints involve the hardware itself, the language used, the problems the programmer is involved with, etc.). The focal level itself is represented by the *dynamic process of finding the limit cycles which terminate the acting out of the programme*. Material causes of this dynamic reside in lower level electrical impulses. Formal causes are found in the Boolean logic of the lower level and in the higher level structure of the programme. Efficient causes are found in the starting of the programme and in any further perturbing inputs, all higher level. Final causes reside in the higher level structure of the lattice, reflecting the world view within which questions are being asked.

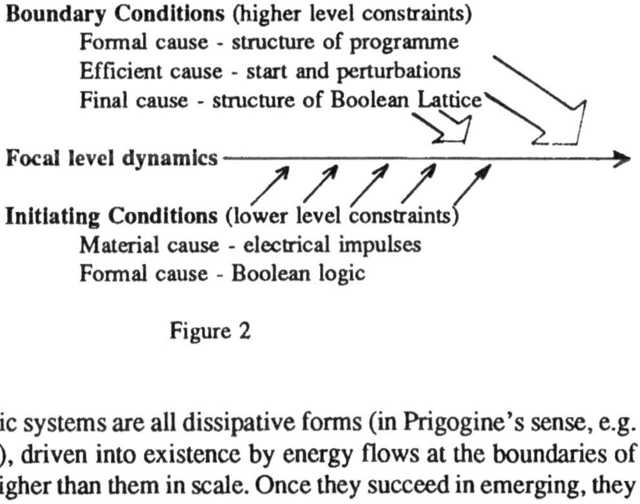

Figure 2

Natural dynamic systems are all dissipative forms (in Prigogine's sense, e.g. 1980; Wicken, 1987), driven into existence by energy flows at the boundaries of other such systems higher than them in scale. Once they succeed in emerging, they undergo a stereotypical *development* from immaturity to maturity, and are than recycled (Table, next page). During this time they traverse a series of characteristic stages. In the early immature phase they appear to maximize the gross energy flow through them (the maximum power principle of Lotka and the Odums) and increase their *ascendancy* (Ulanowicz, 1986) as they *grow*. During their senescent slide into oblivion they are better described as continuing their differentiation through a relaxation period when their entropy production tends to become minimized (Prigogine), although most will be terminated before reaching the minimum entropy production state. In this phase, those with some stability (like organisms) achieve what efficiency is possible to them as they assume increasingly machine-like status. Some primitive systems of this type (e.g. whirlpools) don't seem to grow, others (e.g. dust devils and tornados) never really become stable enough to be described as 'machine-like'.

General Phenomenological Rules of Developing Thermodynamically Open Dynamic Systems

The system shows:

(1) after an initial increase, a monotonic average decrease in intensity of energy flow (i.e. weight-specific energy flow) through it,
(2) a continual, hyperbolic increase in complicatedness (= size + number of kinds of components + organization, in different combinations during different stages of its maturation),
(3) an increase in its internal stability (its rate of development slows down),
(4) a decrease in its stability to perturbations.[1]

Lattices of Boolean automata exhibit development at least in the sense of approaching a relatively small number of alternative limit cycles (Atlan et al., 1986, 1987). That is, their behaviour is predictable within limits, and development can be *defined* as 'predictable irreversible change' (Salthe, *Syst. Res.*, vol. 6). Their behaviour appears to be at least as predictable as ecological succession, for example, which is accepted by all as a development. In addition, if it could be shown that the dynamics of the rate of accumulation of stabilized automata in the network is describable as an S-shaped or hyperbolic growth curve (Figure 3), then they would show some of the typical phenomenology of thermodynamically open systems (in particular, a diminishing rate of accumulation of stored information, or of rules of behaviour). They do not, however, show a diminishing intensity of energy flow through them and thus is precluded the appearance of much of the typical phenomenology shown in the table. Another way they fail to model natural open systems is by not showing an increase in the number of possible states of the system while they are becoming more organized (Brooks and Wiley, 1985 - Figure 4). The lattice does not, in effect, grow while its behaviour specializes.

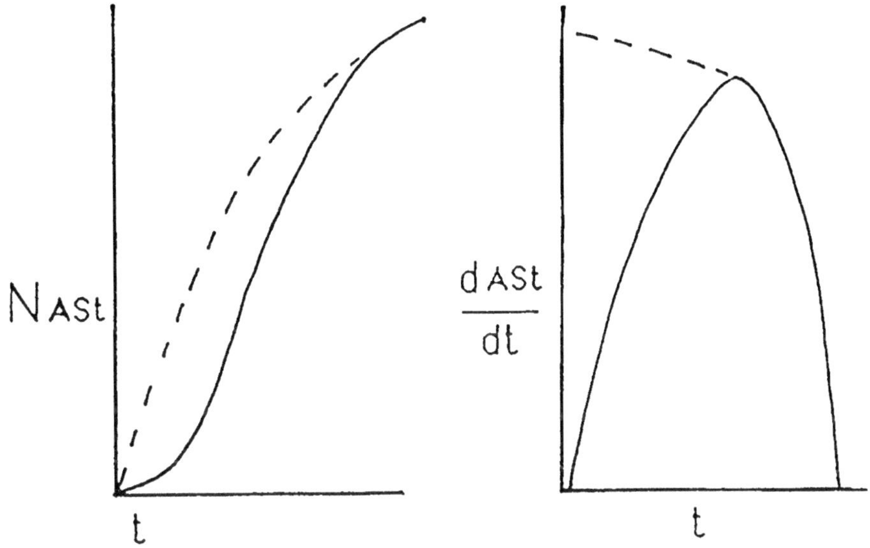

Figure 3

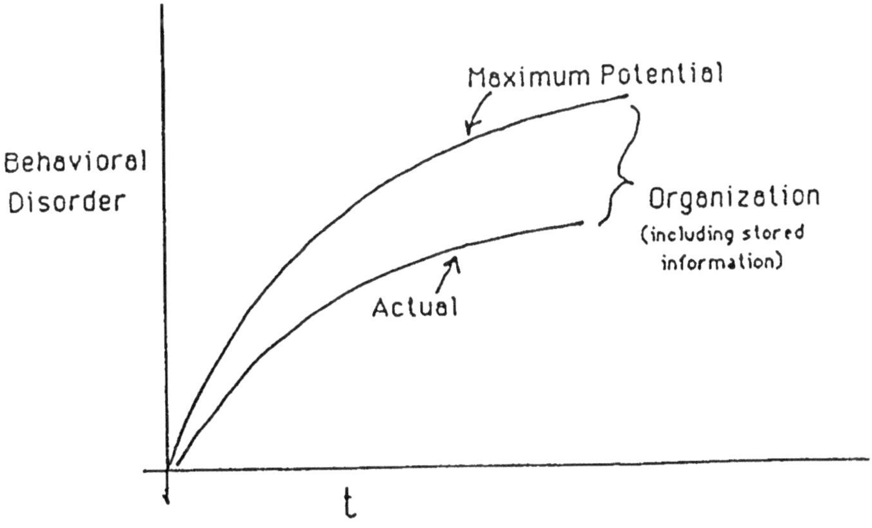

Figure 4

Open systems '*evolve*' in the sense of being irreversibly marked by traces from encounters with perturbations from their environment. Such perturbations may terminate a system if they are of large scale, or if too many impinge during a short period of time, or if they strike some sensitive part of the system. Perturbations survived give rise to altered form in an evolving system. Alterations like these will accumulate as long as the system survives, and contribute to the entropic increase in systemic *stored (macroscopic) information* (this is defined as any combination of numbers of types and tokens, numbers of rules, size) in the system (Layzer, 1977; Brooks and Wiley, 1986). Rules of behaviour can be viewed as the *organization* of such systems, and ascendancy importantly involves an increase in their number. (In primitive systems, rules reduce to friction and inertial forces). In Boolean lattices information is gradually stored as more automata become stabilized, which moves the system away from its maximum possible behavioural disorder as it differentiates (Figure 5). Disorder absolutely *decreases*, rather than increasing as expected from the second law of thermodynamics *because the system cannot grow*, even though energy flow that could be the source of growth *increases* its organization. The system is artificially constrained against growth.

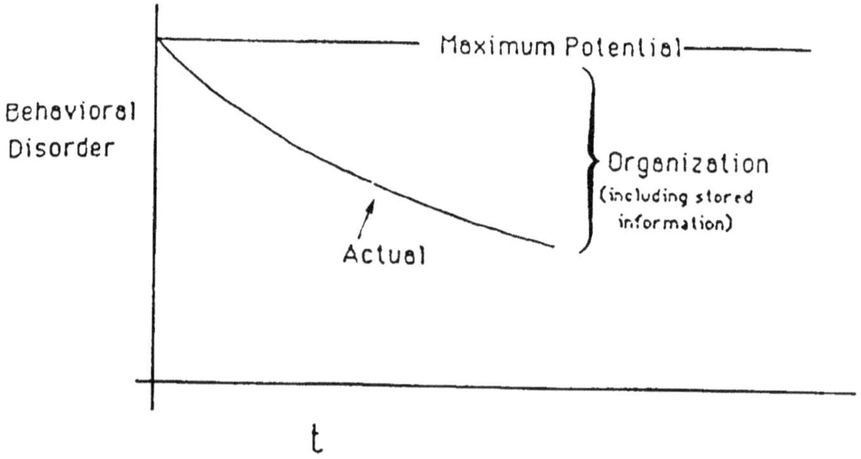

Figure 5

So, Boolean networks do show evolutionary change to the extent that changes of state are not deterministic. Evolution may be defined as 'unpredictable irreversible change'. As I understand it, each separate instance of the development of a given network shows some idiosyncratic behaviour which is preserved in the final states. In addition, the ability to change the final attractor state after perturbation can be taken to be the storage of historical information in the system, hence evolutionary. On the other hand, if the new trajectory is only one of a standard set of them, the change would be developmental after all. In any case, it *does* seem that Boolean networks both develop and evolve — that is, they have some predictability over and above their idiosyncratic behaviour.

System *adaptability* (Conrad, 1983) emerges from the ability to *anticipate* environmental perturbations (see Rosen, 1985). Primitive moves of this kind can be viewed as random with respect to success, relying upon chance coherence or anti-coherence (Wolkowski, 1984) with environmental configurations. More developed systems can be described as following a *project* (*sensu* Sartre), a higher level entrainment, to help them find their way. These projects are traits of the system environment, and to the extent that states of subsystems in that supersystem are correlated, the tracking of one of them by the developing-evolving system can help to anticipate unexpected perturbations from others (Rosen, 1985). Finally, as the system becomes more mechanical, *habits* (*sensu* C.S. Peirce) accumulate which the system follows until its rigidity on this score leads to its recycling, or, as in the Boolean networks, the system settles down to stereotypical behaviour.

Atlan describes Boolean networks as modelling *self-organization*. His definition of this involves (a) a differentiation of the system from homogeneous to heterogeneous based on internal information, while (b) preserving the effects of environmental perturbations met along the way. With that definition, combining aspects of both development and evolution, the lattices are self-organizing, but some important aspects of *natural* self-organizing systems seem to be missing. As noted above, (a) the intensity of energy flow through them does not decrease steadily as they mature, and (b) the number of possible states the system could access does not grow in an early immature phase. Hence, they are frozen into an immature energy flow situation combined with a senescent differentiation process that makes for a *very* unnatural trajectory. As a result they show some peculiar phenomenology as well. For example, as I understand Atlan, perturbations earlier in the trajectory stabilize a smaller number of automata than perturbations later in the trajectory — "The time of application of the [perturbation] is correlated with the size of the set of stabilized nodes" (Atlan et al., 1986). In *natural* self-organizing systems, perturbations later in the trajectory typically have an increasingly *de*-stabilizing effect. I am sure this difference comes from the lack of change of energy flow through the network, or stated otherwise, that the automata do not increase in number with no change in overall voltage, so that each one has less current flowing through it. (I will not do more here than mention the notion that natural self-organizing systems have internalized (focal level) efficient causes, reflecting active agency; I cannot imagine how that could be programmed).

For these reasons, the Boolean networks really are still only cybernetic (even if 'second' cybernetic) rather than truly self-organizing. Cybernetics is characterized by equilibrium dynamics. The system moves toward an equilibrium situation, however tortuously, and however one might perturb it. Cognition in developing-evolving, homeorhetic rather than homeostatic, thermodynamically open systems really cannot be well modelled cybernetically. That will require simulations that can grow during an immature phase, where semiosis will be tied to ascendancy, and that will switch at some point to the senescent trajectory followed by most existing simulations.

Question from Y. Rav: "What is your evidence for supposing the existence of a 'mature' stage of development in natural dynamic systems of all kinds ?"

A: "There is no evidence as such, but a chain of permissible reasoning, as follows. We know that we, as mature individuals, are not intelligibly modelled as simply moving from what I have called 'immature' to 'senescent' states. We know that we are agents. Yet aggressive scientific hypotheses make that move, just as they have made it for natural systems in general. If they can be so wrong about us, then they might, as an hypothesis, be just as wrong about other, perhaps *all*, natural systems. Perhaps we are missing some important aspects of nature by fiat of assumption. It is certainly believable that pets, for example, might for a while be agents, which extends the hypothesis plausibly to animals. Having gone that far, and knowing that there is no possibility of testing the notion, why not expand the condition to *all* dissipative forms. The machine hypothesis was arbitrarily so extended throughout nature by an equally untestable assumption. Having done this extension, we can reasonably pull back to the extent of admitting that the mature phase of, say, a hurricane, might last for only a minute portion of its development, or that in, say, modular plants, the immature phase tends to be extended proportionally far beyond what it is in many other kinds of dissipative forms. The generality then becomes the immature → mature → senescent trajectory, with deformations characteristic of different kinds of systems, and, perhaps, even losses of some stages in some systems."

* This article was first published in *Communication & Cognition*, vol. 23, nos. 2-3.

Notes

1. From data in Trincher, 1965; Margalef, 1968; Odum, 1969; Zotin, 1972; May, 1973; Wesley, 1974; Brooks and Wiley, 1986)

References

Atlan, H.; Ben-Ezra, E.; Fogelman-Soulié, F.; Pellegrin, D.; Weisbuch, G., 1986, Emergence of classification procedures in automata networks as a model for functional self-organization, *Journal of Theoretical Biology*, no. 120, pp. 371-380.

Atlan, H., 1987, Self-creation of Meaning, *Physica Scripta*, vol. 36, pp. 563-576.

Brooks, D.; Wiley, E.D., 1985, *Entropy as Evolution*, University of Chicago Press.

Conrad, M., 1983 *Adaptability: The Significance of Variability from Molecule to Ecosystem*, Plenum Press, New York.

Layzer, D., 1977, Information in Cosmology, Physics and Biology, *International Journal of Quant. Chem.*, vol. 12, suppl. 1.1, pp. 185-195.

Prigogine, I, 1980, *From Being to Becoming: Time and Complexity in the Physical Sciences*, Freeman, San Francisco.

Rosen, R., 1985, *Anticipatory Systems*, Plenum Press, New York.

Salthe, S., 1985, *Evolving Hierarchical Systems*, Columbia University Press, Columbia.

Salthe, S., Self-organization of/in hierarchically structured systems, *Systems Research*, vol. 6, pp. 199-208.

Ulanowicz, R.E., A phenomenology of evolving networks, *Systems Research*, vol. 6, pp. 204-217.

Ulanowicz, R.E., 1986, *Growth and Development*, Springer Verlag, Berlin.

Wicken, J.S., 1987, *Evolution, Information and Thermodynamics*, Oxford U. Press.

Wolkowski, 1984, *Synergie et Cohérence dans les Systèmes Biologiques*, E.E.E.E.

A PRIORI AND A POSTERIORI IN COGNITIVE PRAXIS THE MODEL FOR THE REGULATION OF AGONISTIC ANTAGONISTIC COUPLES

E. Bernard-Weil

Clinique Neuro-Chirurgicale de l'Hôpital de la Pitié
Paris

Abstract

Three couples have to be taken into account to advance in cognition researches: the couples formed by a priori and a posteriori, *praxis* and cognition, by model of cognition and model of the known, to which may be added the couple of realism and conceptualism (or conventionalism). A general model, called the model for the regulation of agonistic antagonistic couples (MRAAC) allowed us to better accept and even to simulate the dynamics of these couples without having to establish a hierarchy between the two elements of the couple or without having to search a middle way. We emphasize, for that matter, the similarities between the 'model' of the genetic epistemology (GE) - which is a model for cognitive activities - and the MRAAC - which is both a model of biological as it is for cultural systems. Such a type of cognitive praxis would give new possibilities of knowledge and action as far as bilateral strategies are concerned which are perhaps the sole strategies capable of avoiding the so-called 'perverse' effects of some unilateral strategies. As far as other epistemological 'applications' are concerned, we show how this model permits new interpretations of diverse philosophical or theological works. Finally, a few examples are evoked, demonstrating the fruitfulness of considering the circularity, the paradox, the double-bind, the agonistic antagonistic couple as a foundation (but how evanescent!) of cognition and praxis.

THE THREE COUPLES

1. The cognition - praxis couple

What does this breaking of the psychological and neurophysiological unity

G. Van de Vijver (ed.), New Perspectives on Cybernetics, 59–76.

of the living being mean to the physician ? At first sight one could believe that my professional activity makes use of a cognitive track to begin with and that this would justify a treatise only about this one track, at least as a first step. *Then* such a cognitive way would lead to the diagnosis, and, by referring to the corresponding chapter on the book of therapies, we subsequently could depart from this diagnosis to elaborate the convenient treatment.

Now this is not true. To a physician educated as much in the field of clinical sciences as in that of nomothetic sciences, the diagnosis after all is not an end in itself, the important thing is to perform a correct action that complies with his technical and social vocation. Diagnosis is only a guide-mark, a caricature of reality, a kind of 'degeneracy' due to the lack of complete information concerning the patient and his environment and due to the more or less loose construct made with that sort of information. This is another way of saying: "there are no diseases, there are only patients". Moreover, the phase of diagnosis is all the time superseded by the retroaction of its own future, i.e. the praxis, in the same way that the meaning of a sentence, according to Lacan, and also its performance value, appears only when the sentence is concluded.

On top of that we have noticed with regard to the researches of Simon (1986), that he could not wholly maintain the dogma or the central idea of the 'information treatment' in his definition of 'artificial intelligence' (AI), and that he had to propose to add the concept of 'motivation' to it. Between motivation and praxis there are not many differences except that the latter is virtually included in the first. Here though the destroyed couple cognition/praxis (namely by suppression of the second element) was re-established.

Finalities associated with cognition and the role of motricity in the same phenomenon gave rise to many neurophysiological or neuropsychological works. Our aim is not to recall them here. What we aim at is to recall the existence of a type of modelling, combining praxis and cognition, that illustrates our thesis (cf. *infra*).

But first we want to end this account concerning the general characteristics of the couple cognition/praxis. Consequently we will relate other couples to it which belong to the same family.

As far as this couple is concerned, we would like to show that we do not want to go in for a complete reversal in the cognitive researches, of which we accept the more recent conclusions.

To the left of Figure 1, one can see the cognitive activity that begins beyond the so-called 'operational closure' (OC) (Varela, 1987) or '*interface*' (Tabary, 1986). We do not object to the term of OC, although we are not convinced that there exists a radical difference between the coupling by inputs of cybernetics and the coupling by perturbations of systemics (or second order cybernetics). We have to take into account the way cognitive autonomy is conceived. In Varela (1986) this autonomy needed the OC in order to be maintained: OC was then defined by permanent interactions between the sub-systems that formed it according to a rather traditional 'mechanics', even if this internal coherence, these eigenbehaviours could be metaphorically represented by iterate equations. In Tabary (1986) the definition of the autonomy, although chronologically later, seemed more valuable, insofar as the ideas of Edelman, Mouncastle, Jerne concerning the processes of degeneracy and selection, were transposed towards such topics. Nevertheless, Tabary was brought by these facts to reject every form of realism. According to this author we perceive or know only the *interface*, in the same manner as we are operationally closed according to Varela. So they both brought some arguments to bear to sustain the thesis of the relativity of knowledge or in other words of the nominalist point of view.

But the problem becomes much clearer if one pursues it beyond the cognition in a strict sense. Cognition gives rise to a praxis, effective or only in outline, that directly tackles reality and produces an inversion of the degeneracy of perception and information towards a kind of 'regeneration' where our actions are able to adjust themselves to the constraints of the real and to change a decision of conjectural origin in an objective effect upon the circumjacent world. With Vallée (1984) one can find once more such a return to the perception of a real transformed by the shock of the decision, allowing the loop to be never disrupted, rhythmed in a certain way by the decisions and the successive perceptions. Even with Tabary such a paradigm seemed to be outlined. It could, according to him, renew the bases of AI, instead of (or in addition to), the data-base, the rules and the inference motor.

2. *The couple cognitive model/model of the known*

In this connection, we have to underline that the mechanisms usually proposed for expert systems do not seem to correspond, in our opinion, to the medical expert activity. The physician could (in)directly come into contact with the real of the patient, without the mechanisms of degeneracy and selection intervening, because of the existence of another couple. This couple is formed by the internal model of cognition, the model that one could identify in the known system and that would function ... on the basis of the same model. To give just one example, the equilibration processes, mainly between assimilation and accommodation, described by Piaget (1975), do not seem fundamentally different from the equilibration processes we observed between both elements of a couple of hormones, the couple cortisone/vasopressin or insulin/glucagon, or the innumerable couples recognized by molecular biology (MB) -- combining metabolic stimulation and inhibition, gene expression and repression -- of all which MB has not yet discovered the epistemological and practical consequences. In the following we will develop the very suggestive parallelism between the 'model' of the genetic epistemology and the 'model for the regulation of agonistic antagonistic couples' (MRAAC), derived from a bio-medical experience.

3. The couple a priori/a posteriori

Before we give a full account of the model that will conclude this paper, it is worth noting that, in order to make its approach easier, we have evoked a series of couples, the last one being the couple a priori/a posteriori.

The position of this couple in this paper is self-referential. Let us explain this point: when we wrote the summary of our communication that had to be presented to the members of this Colloquium, we had forgotten the exact title we previously proposed. And when we recalled this title, we tried to find back in the summary the idea included in the title. The relation between the title and the summary was not immediately evident! As a matter of fact, from a title given 'a priori' we progressively discovered 'a posteriori' that it was fairly well described in the summary, but filigreed so to speak. Even if this example is not absolutely convincing, we are quite observant of everything which may appear as 'a priori' in the cognitive activities of the sciences which are called experimental or inductive, although this type of researches does not seem to be accepted easily by the scientific community dealing with it.

Let us take up again the example of the genetic epistemology. Cognition is supposed to be constructed and not innate. The newly-born would dispose of only a few innate activities (sucking, ...) that afterwards though proved to be more numerous and complex though than Piaget thought. A very important innate activity and one which concerns the question raised here, is the existence of an equilibration process, 'preceding' every construction. *Never could the equilibration process be considered as built by the cognitive activity* in the same manner as the latter was able to elaborate the intuitive and operative schemes studied by the genetic epistemology. Piaget rather tardily asked himself what was the nature of this process. He wrote (Piaget, 1975): "We understand this not as the application, to whatever situation and at every level, of one and the same general structure of equilibrium, given once and for all, as is the *Gestalt* (inspired by the rules of the 'field') for the psychology of form, but as a process (hence the term of 'equilibration') leading from some states of approximate balance to other, qualitatively different ones, passing through multiple imbalances and re-equilibrations".

What we want to say is that there is at the basis or at the origin of each knowledge a circularity, a paradox that, because it is a paradox, causes the notion of basis or origin to disappear.

Here are three other examples, known already or appearing here for the first time. The DNA, according to the genetic code, allows for the production of proteins by the cell; but, in order to read the DNA code, a first protein has to be present already, able to decipher the formula of the protein to be synthesized. It would seem that the biologists have not brought a satisfactory response to this disquieting aporia (Hagène et al., 1986).

The paradox of the autonomy was dealt with by Vullierme in a thesis on the concept of political system (Vullierme, 1990): he 'isolated' inside the society a political structure which he called 'autonomous community' (AC), but this was autonomous only as far as the *oikoi* (home, trade, religious groups ...) — which constitute the society and whose AC guarantees the respective autonomies, were already autonomous, or, more simply, as far as the society, which can only be founded by an AC, pre-exists to the AC. Therefore, AC, juste like DNA, needed its own products in order to act.

The *Urverdrängung*, our third example, was apparently not seen as eliciting a problem of paradox or circularity. Schematically, every repression spoken about in psychoanalysis bears some connection to the refusal of consciousness to take up what is called the pulsional representation, but also with an attraction issuing from the already unconscious contents - from which derives the necessity of an *Urverdrängung* or primal repression that could incorporate new unconscious formations. But what attracts the primal repression ? As no one seems to have been disturbed by such a question up to now, we will not sound the alarm, although formally, this paradox is quite close to the preceding ones.

Where there is no primal or originated paradox, there is, in our opinion, no valuable rationality. So we agree with Castoriadis when he talks of 'the undefinable in the definition of the defined, the ineliminable circularity of every undertaking directed at foundation' (Castoriadis, 1983). And the paradox redoubles, as we have already said, within the notion itself of the primal or originated paradox, because the term 'primal' could make us believe that we are confronted with a first notion from which it would be possible to build up (or take into account) the ensuing rationality. However, this is not the case at all: the paradox is not *introduced* by us, but is given to us, even *revealed* we could say, if the latter word were not to evoke some unpleasant connotations. This paradox reminds us somewhat of the meteorite shown at the beginning of Kubrick's Space Odyssey', or we can see it as a starting-box we don't encounter again during the race, except when we are launching ourselves for a new record.

This is the reason why we will say that every theory of cognition should be based as much on the a priori as on the a posteriori, *thereby giving up all hope that the a priori could ever be justified by the a posteriori.*

4. Delusion of the 'Middle Way'

If it seems impossible to choose between the praxis or the cognition a preferential term from which the other would be deduced, or for that matter between cognitive model and model of the known, or between a priori and a posteriori, it is no longer allowed, in our opinion, to look for a way of escape by means of a hypothetical third way or middle way. It is a temptation for those who cannot endure such a mental 'quartering' or cannot accept what they consider as an abdication of the reason, or who cannot adhere to a kind of epistemological faith that is finally a submission to the 'agonistic antagonistic' law: a law that, once accepted or to be more exact: found again wherever it legislates, could give rise to very important changes in social and biological organizations.

In other investigations (Bernard-Weil, 1985b, 1987), we have given other examples of such 'collapses' (i.e. the impossibility to stand a couple with two equal and coexisting elements), both out of the platonism rejecting the discoveries of the Presocratics, as out of the bulk of the expressions of the 'pensée sauvage' as opposed to other cultures where the notion of couple excluding a third way met with the most perfect expression ("Moving by making to climb, moving by making to descend, that is the life of the world" as said to Griaule by his informant Ogotommeli (Griaule, 1966)). Now one may see with some general systems searchers, who in other respects made a valuable contribution to second order cybernetics, a reference to a medieval Indian school, called Nagarjuna. This school of the Middle Way would go beyond the dialectic opposition between objectivity and relativism (or conventionalism) (Varela, 1987). But our scheme (Fig. 1) showed there was no other 'middle way' than to follow in turns the opposite ways of relativism and objectivism. Tracing a tradition back to its source, that of the Rig-Veda, will show some dynamical structures which approximate this conclusion far better than the Middle Way did. Such comments will be better understood if we transfer them to the end of this paper, after having given a view of the MRAAC.

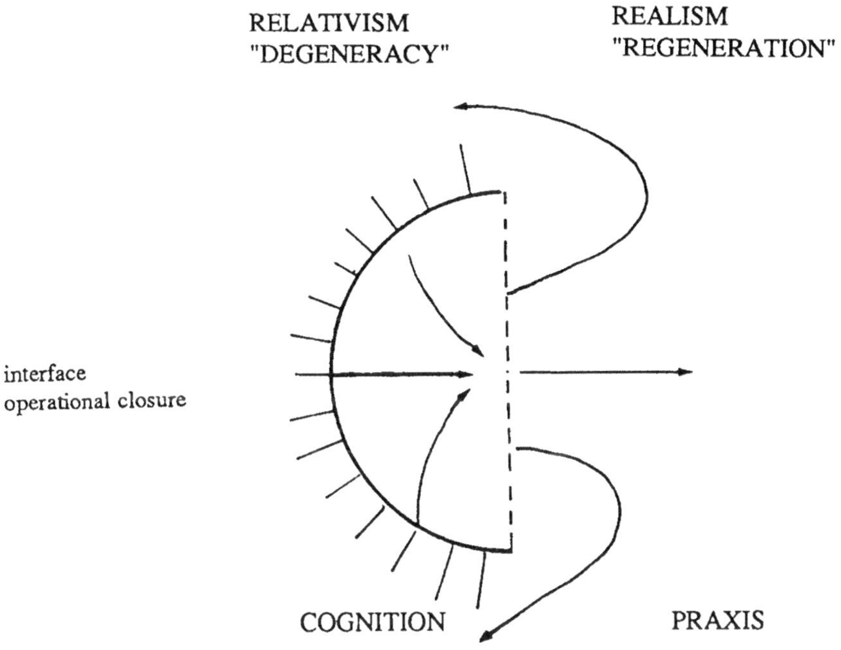

Figure 1

Cognition and praxis

THE MODEL FOR THE REGULATION OF AGONISTIC ANTAGONISTIC COUPLES (MRAAC)

This model has been built from a bio-medical experience. Such an origin is contingent in so far as it might have been built from a very different experience, seeing that it was destined to be extended to other fields than the one where it was originally conceived. It can be considered as a new version of a very old model already existing in the Bible, the Presocratics, Saint-Bonaventure, Moïse of Leon, Pascal, Hegel, Nietzsche, Freud, Lacan, Heidegger, Thom, Girard, Piaget, ... and which the structuralism has caused to be taken up again, but in a static form

(Bernard-Weil, 1987). This model can only be understood from a 'mimetic *poiesis*' (Nancy, 1985) or a '*danse des signifiants*' (Deleuze, 1969), i.e. by making it work at the same time inside our cognitive activity and in the observed system. This very particular pedagogy resorts to various metaphors, of which we will point out only one, that of the '*flying scaffolding*': if both ropes have the same length, the scaffolding is horizontal (=security); if the ropes have a total length equal to twice the distance between the roof and the place where the scaffolding has to be placed for a given work, it is at the convenient level (=efficiency); a computer on the roof controls two winches, which maintain the scaffolding in a good position, whatever the perturbations. The antagonism (or the 'balance') corresponds to the horizontality, the agonism (growth - decrease) corresponds to the height of the scaffolding. If, in spite of everything, an imbalance is present (an inequality of the height of the extremities for instance), the computer that is out of order will try to prevent an attempt to undo the imbalance: if we were to attempt to lower the one extremity that has gone up, the computer would provoke a release of the rope tied to the other extremity ... The 'therapeutical' strategy may seem astonishing. It will bear on both extremities, the one that is too high *and* the one that is too low, with unequal and variable forces. A mathematical model (Bernard-Weil, 1985a, 1988) allowed us to simulate these diverse occurrences (Fig. 2 and 3).

Nevertheless, we can also propose another discourse about this model, of a 'university' type that may be considered as complementary to the first one which is of a holistic or systemic type. Here are its characteristics, which account for the properties of the three couples which were used to introduce it.

1) its definition: the two elements of the couple act, as we have seen, from an antagonistic point of view (ANT), i.e. with some actions of an opposite sense on some receptors, and from an agonistic point of view (AG), i.e. with some actions of the same sense on other receptors. To each of these types of actions correspond some regulations, crossed for ANT, parallel for AG, that have to fulfil some norms: the ANT norm, which fixes the difference between the values of the opposite forces (relative values), is in general equal to zero, but may correspond to a non-null value in some systems ($x(t)-y(t)=a$, for instance); the AG norm, which fixes the sum of these values (absolute values), corresponds to a non-null value, constant or variable in relation to time ($x(t)+y(t)=m(t)$, for instance). The 'computer' connected with each couple *allows to re-establish the ANT and AG balances after a perturbation, except in the 'pathological' cases where it endeavours to reestablish the imbalances after perturbation*, namely after a perturbation produced by an inappropriate therapeutical act or by control (unilateral strategy).

As the model is 'non-linear', it functions in general in the form of cycles or oscillations. If the model functions badly, the critical points of the imbalances are multiple, none of them fulfilling the norms.

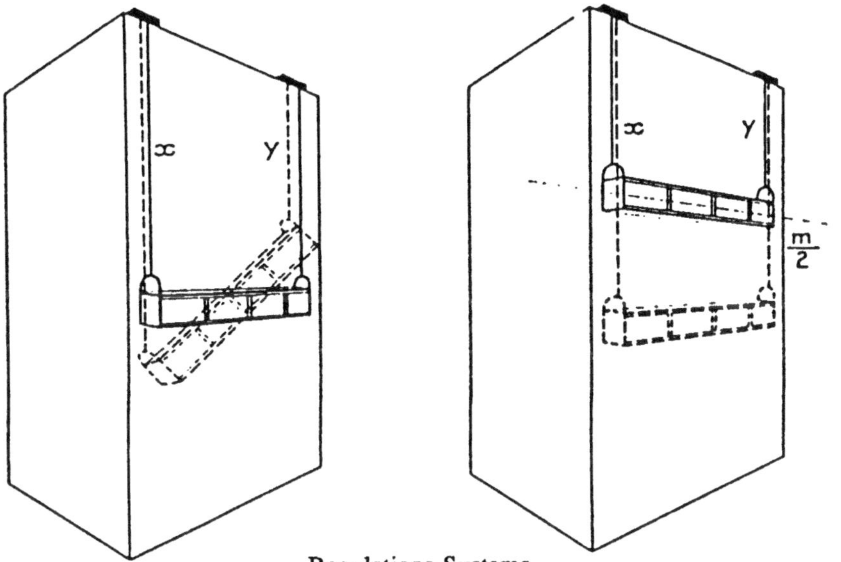

Regulations Systems

crossed
x - y → 0

parallel
x + y → m

Figure 2

The Flying Scaffolding

We may also represent the four poles of the model in the following way: the two forces at the two extremities of a horizontal axis, on a vertical line passing through the middle of the previous one, at the top we put the 'computer' and at the bottom the receptor of the forces.

The similarities of this model with the Piagetian model are evident: in the vocabulary of genetic epistemology, the processes of compensation (under the two forms of inversion and reciprocity) correspond to the ANT regulation, whereas the processes of construction (*equilibration majorante*) correspond to the AG regulation. Now we find such a dynamics also in the systems (biological ones for instance) that our cognitive activity deals with.

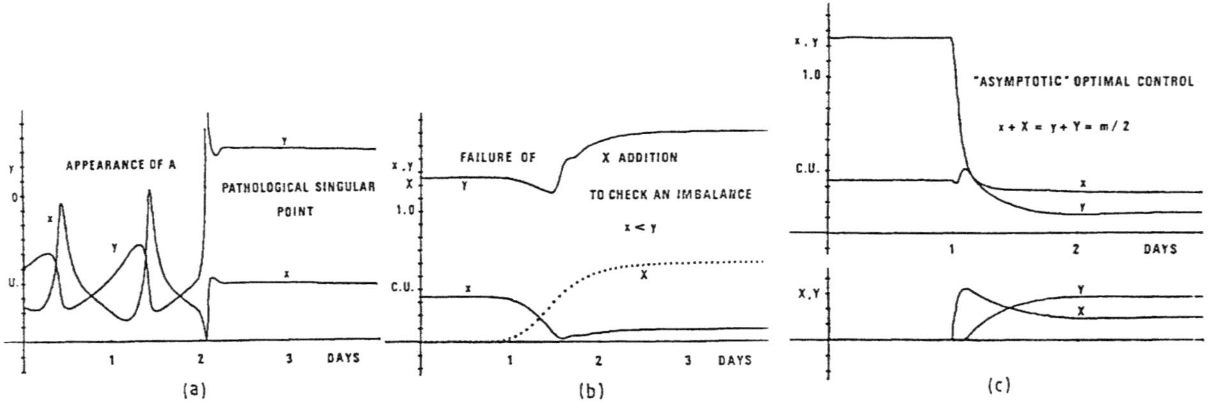

Figure 3

Examples of simulation with the mathematical version of the Model for the Regulation of Agonistic Antagonistic Couples. At the top: a) transition from physiological circadian rhythms of the cortisol (x) and vasopressin (y) to an asymptotical imbalance x<y; b) inefficacy of the addition of cortisol (X); c) a solution given by the model: to administrate together X and vasopressin (Y). Some therapies took pattern from these results (and from other curves tending to reestablish the circadian rhythms).

2) As far as the thermodynamics of Prigogine are concerned, the agonistic antagonistic (AA) processes are far from the equilibrium in the thermodynamical sense. But the specificity of the MRAAC lies in the fact that the systems 'far from equilibrium' may be either in a state of balance or in a state of imbalance!

3) The notion of *constituent division* (*division constituante*), a term borrowed from Jacques Lacan, represents one of the major difficulties for the access to the MRAAC. We must accept an absolute severance (an '*abîme*') between both elements of the couple, and at the same time we must accept that they belong tho the same formal and organic unity. The '*manque*' (lack) cannot be sutured, but we can go beyond it (temporarily) if the two elements that it separates correspond to a balanced model - in some events such as the Lacanian 'bonne fortune' (*tuchè* in Greek), or the 'epiphany' with James Joyce, or the 'privileged moments' of Jean-Paul Sartre, or the good health in the domain of the biological couples, or the phallic enjoyment in a psychoanalytic sense.

4) The '*super-model*', now called AA networks (Bernard-Weil, 1991), associates several 'elementary' models, but always in an AA form. It balances, in an AA manner, two of these models, or more. So, we find at all the 'levels' the same AA structure, which allows, in our opinion, to reject or at least to reconsider the classical notions of hierarchic systems, levels and meta-models (see *infra*). Such a conception is close to the fractal and holographic theories of cognition (Pinson et al., 1985; Tabary, 1987), seeing that each element is an image of the whole, and the other way round. (We would like to emphasize the fact that these theories were inspired by other researches concerning the holographic point of view for memory (Pribram, 1977) and physical order (Bohm, 1980)). Piaget was not so far from this mode of conception when he described the three equilibria: "... the equilibrium of the relations between the subject and the objects, essential as far as the physical and experimental knowledges are concerned, the equilibrium of the coordinations between schemes or between sub-systems of schemes, which procures the logico-mathematical knowledges, and the general equilibrium between the whole and the parts, and therefore between the differentiation of the schemes or sub-systems and their inclusion in a complete system" (Piaget, 1976).

5) *The dichotomies of the modelling*: the MRAAC assumes the dichotomies, without making their synthesis, but expressing in turn the opposite aspects. Let us enumerate these dichotomies: a) open/closed; b) synchronic/diachronic; c)homeostatic/homeorhetic; d) local/global; e) general model/concrete model; f) hierarchy/autonomy; g) a priori/a posteriori; h) model of the knowledge/model of

the known; i) simplicity/complexity; j) empirico-intuitive model/formal model; k) theoretical/practical.

6) *The pathological homeostasis*: We are dealing here with an abnormal control (in relation to the norms of the model), due to a change in the 'parametric field' of the model, around pathological critical points (cf. 1).

7) and 8): the two last characteristics elicited much controversy (contrarily to the preceding ones) during our seminar at the International College of Philosophy in 1987. They do not seem to interfere directly with the content of the present paper, but it would be better not to elude them completely. The seventh characteristic concerns the existence of *false AA couples*. Although considered as couples of opposites from a semantic point of view, the following couples are not AA couples (cf. Bernard-Weil, 1988): good-evil, order-disorder, perhaps life-death, finally the One (in the presocratic sense)-the Multiple. The eighth characteristic concerns the *meta-model of a general model*, of MRAAC type or similar, which is not constituted in itself as a hierarchy of models and meta-models. A meta-model of a general model is logically necessary, but it is impossible to say anything about it with the help of this model, even if the model would be general or (quasi-)universal, because the meta-model is the condition of the appearance of this model. To us, the meta-model, ill-suited to be modelled, is the sphere of freedom, of creation and innovation - the model itself ensuring nothing but the management (for instance, the AA couples are formed in the meta-model, 'and then' they are managed or made viable by means of the MRAAC). In the general model of Hegel, the philosopher has missed out the meta-model (Derrida, 1981). The meta-model was present in the trinitarian model of Saint-Bonaventure or in the sephirotic model of the *Zohar* (Bernard- Weil, 1985b, 1988).

APPLICATIONS OF THE MRAAC

Although applications sometimes occur before the theory, we will stick for convenience to the term 'applications'.

1. In biomedicine

The MRAAC allowed us: a) to reconsider the system formed by the adrenals and the neuro-postpituitary; b) to identify the first polypeptidic cell growth factor, as such as early as 1968, namely, the vasopressin (post-pituitary hormone), (Bernard-Weil & DaLage, 1968) (there was an empty part in the table on the ANT actions of these hormones, with regard to the 'anti-growth factor' effects of the cortisone (adrenal hormone); c) to start with the further development of the so-called bipolar therapies in case of an imbalance between the endogenous secretions of these hormones, by simultaneously resorting to the 'good' hormone, the one that is lacking, and to the 'bad' hormone, the one in excess, whereas the use of only the first one always resulted in a failure; it is an example of *bilateral cognitive praxis*, the extended use of which seems desirable to us *wherever the unilateral praxis, i.e. not coupled with a bilateral cognitive praxis, elicits some 'perverse' (or counter-intuitive) effects.*

2. In the history of philosophia and theologia

The MRAAC represents a '*grille de lecture*' that allowed us to find once more some dynamical structures, close to this model, inside a family of philosophical statements or of theological revealings (Bernard-Weil, 1975, 1984, 1988).

As a new example, let us evoke the Indian myths that could be dependant on such a modelling. They surely do not correspond to the Middle Way, as we said already, or more exactly, they do not correspond to a certain understanding of the Middle Way (cf. infra)). We retain only two 'applications'.

a) the three *gunnas* are three qualities of the matter that give rise to the five elements by their 'connection' and they formed the subject of a lot of developments in the *Bhagavat-Purana* (probably composed towards the 4th century A.C.): *sattwa, rajas* and *tamas* are very polysemantic vocables; by choosing the translations that are convenient for our modelling, one finds respectively, harmony, energy, inertia. The two last terms form an AA couple, and the first may be considered as the regulator (or the equilibrator ... or the control computer) of these opposite effects, although such a dissociation of these two orders or 'levels' was not always evident in these rather late texts.

b) The *Rig-Veda*, that is older (twenty centuries before), allows for an interpretation that is even more to the point. In the *Hymn to Agni* (I, 95), one clearly sees the triple partition of the god: firstly a couple formed by *Hari*, fire of the sacrifice in the temple, and *Soucra*, the fire of the sun. They belong to the same conceptual scheme because they were "born from two mothers of different colours". On the other hand, the third name of Agni was *Twachtri*, who has an intermediary position, being the fire of the storm and of the clouds. He corresponds to another order than the previous ones because he was not born from a mother but "came out of the bosom of *Arani*, aided by ten young ministers" (out of the Log from which springs up the flame produced by the ten fingers) (Langlois, 1848). This notion of duality, organized but not suppressed by the mediator, was found again in other hymns to the god *Agni*, considered a unity in relation to new couples this time: "His beam, spreading over with greatness, has warmed the Earth and the Heaven" (X, 88) or "He holds out his arms, and, terrible, he exerts himself to form his double dress" (I, 95).

It may seem that this analysis has not been too much impoverishing, as all explanation of a text risks being. According to Leturmy (1958), the vedic Hymns are not "a dogmatic treatise, but must be left in their ritual context of worship". Behind the triple partition we have performed, one will perceive - we hope - this underlying dynamics that echoes back the statements of a poet-philosopher Deguy (1982): for him, the veritable human being - and the authentic scientist to us - has to leave the 'etiological desire' (the desire of finding a cause) for the 'desire of admiration' - and of action, let us add. We must also point out that the Middle Way of Nagarjuna should perhaps be considered from the same point of view. A Tibetan poem said: "Give me your grace in order that I understand the deep view of Nagarjuna: the non-contradictory and mutually beneficial function of the two levels of truth (*samsara* and *nirvana*)" (Yéshé & Zopa, 1985), a sentence badly in need of a synthesis of the two terms.

3. *In the epistemology of cognition*

The MRAAC may help us to conceive the unavoidable duality (with its 'constituent division' and its possibilities of 'equilibration') and to understand the expressions of this duality in our world (by including only the biological and cultural (phenomenal) for the time being). We should like to point out that this sentence may be considered as a paraphrase of a sentence of Parmenide (*Krè to legein te noein t'eon emmenai*: It is required to tell and to think the essence of being (Sauvage, 1983).

Thus the MRAAC represents a pedagogical tool for recognizing the circularity, the paradox, the 'double bind' as a starting point and a point of arrival of every thought. Yet, they are virtual points because they are not easily severable from the concrete phenomena of which they ensure the functioning, or rather, the viability.

The hypothesis, not demonstrable and not falsifiable, of a 'computer' controlling the balance of the AA couples, is also very useful to accept the co-existence of severance and continuity. The MRAAC may be put together with the Lacanian topological models, which tried to make us understand the continuity of the *Umwelt* and the *Innenwelt* - though kept apart by the 'lack', equivalent in this example to the operational closure or the *interface* in cognition processes. *Without such a hypothesis, the bilateral strategies or therapies would not have appeared,* because they were based on the idea of a dysfunction of this 'computer'.

The dichotomies of the AA modelling represent, themselves, a great part of the epistemological challenges of our time. We have considered three among them. Here are some other developments.

If the 'local' is sometimes the consequence of the 'general', we also have to conceive the general from the local (or the absolute *Grund* (reason, basis) from the definite *Grund*, as Hegel formulated this sequence), or even a general model from a local model. Without such a proceeding, the 'general' is doomed to spread out like a fog above the deserted heath of our mind. Let us, in passing, report a remark of Stengers at the Colloquium of Cerisy about 'Auto-Organization' (1985), the meaning of which ran along the lines of this: if biology is looking for some renewal from a general epistemology (after trying to get this from the techniques), this behaviour could be sterilizing in course of time because the current must run in the two opposite directions. A very concrete example is needed to show that the respective roles of the a priori and the a posteriori (a couple that is different from the couple general/local but which may help us in understanding the problem better) cannot be isolated or hierarchized in the case of our medical research. Surely a logician or a searcher in cognition would have been able to discover a priori the MRAAC, without referring to clinical or experimental researches, seeing that this model corresponds to a new 'category' (in its Kantian sense) or better to a 'maxim' (term used by Kant in his *Kritik der Urteilskraft*) such as: *each element of a couple varies in the course of time in relation to the states of ANT and AG balance or imbalance of the system to which it belongs* (it should be noticed that, in spite of the apparent complexity of the mathematical equations of the model, there was nothing else at the basis of the formalized model). Nevertheless, more than fifteen years of clinical observations and experimental investigations were necessary in order that this 'maxim' could be stated.

The reasoning would be the same for the couple *theory/praxis* because the second element of this couple has led the theory towards some frontiers that the sole observation of the phenomena might not have reached, even if the praxis took advantage in its turn of the theoretical contribution.

As to the couple *cognitive model/model of the known*, it seems rightful to assert the homomorphism of the two elements, beyond the conflict that determines their relations. Here we are dealing too with a hypothesis that is neither demonstrable nor falsifiable, although our repeated comparisons during this communication between genetic epistemology and MRAAC were very suggestive. Moreover, this hypothesis could be adopted in other fields than the field where it 'appeared'. In fact, it was not a hypothesis but the condition itself that allowed us to construct the MRAAC.

Finally, a lot of impediments which stand in the way of attaining a true knowledge and a correct action - truth and justice are nowadays words which are little used by men of science, separately they may invite smiling, but together they seem to be endowed with an irresistible force - could be overcome, circumvented or deleted according as to whether the acquirements of the so-called agonistic antagonistic *phylum* (which the MRAAC is an example of, among others) will obtain an increasingly important place in the field of the cognitive epistemology and of the praxeology.

References

Bernard-Weil, E., 1975, *L'Arc et la Corde*, Maloine, Paris.

Bernard-Weil, E., 1985a, Lack of response to a drug: a system theory approach, *Kybernetes*, vol. 14, pp. 25-30.

Bernard-Weil, E., 1985b, Théologie et Systémique, in *Violence et Vérité*, P. Dumouchel - (ed.), Grasset, Paris, pp. 93-109.

Bernard-Weil, E., 1986, La systémique ago-antagoniste en tant que Post-structuralisme, *Proc. du 11ème Congrès International de Cybernétique*, Namur, Association Internationale de Cybernétique, pp. 311-322.

Bernard-Weil, E., 1987, De la modélisation comme essence de la technique, *Revue Internationale de Systémique*, vol. 1, pp. 23-36.

Bernard-Weil, E., 1988, *Précis de Systémique Ago-Antagoniste. Introduction aux Stratégies Bilatérales*, Limonest, L'Interdisciplinaire.

Bernard-Weil, E., 1991, Contrôle d'un système complexe déséquilibré par action au niveau d'un seul "sous-système". Réseaux ago-antagonistes et attracteurs étranges, *Revue Internationale de Systémique*, vol. 5, pp. 27-52.

Bernard-Weil, E.; DaLage C., 1968, Inhibition by cortisol of the favourable effects of lysine-vasopressin on the growth of HeLa cell cultures, *Experientia*, vol. 24, p. 1001.

Bohm, D., 1981, *Wholeness and the Implicate Order*, Routledge and Kegan Paul, London.

Castoriadis, C., 1983, La logique des Magmas et la Question de l'Autonomie, in *L'Auto-Organisation*, J.P. Dupuy and J. Dumouchel (eds.), Seuil, Paris, pp. 421-443.

Deguy, M., 1982, Onglets de lecture, in *René Girard et le Problème du Mal*, M. Deguy and J.P. Dupuy (eds.), Grasset, Paris, pp. 87-61.

Deleuze, G., 1969, *La Logique du Sens*, Ed. de Minuit, Paris.

Derrida, J., 1981, *Glas*, Denoël/Gonthier, Paris.

Griaule, M., 1966, *Dieu d'Eau*, Fayard, Paris.

Hagène, B.; Lenay, C.; Locquin, M., 1986, *Aux Origines de la Vie*, Hachette/Fondation Diderot, Paris.

Langlois, A., 1848, *Le Rig Veda*, Paris.

Leturmy, M., 1958, *Dieux, Héros et Mythes*, Club Français du Livre, Paris.

Nancy, J.L., 1985, Dies irae, in *Faculté de Jugement*, J.F. Lyotard (ed.), Ed. de Minuit, Paris, pp. 9-54.

Piaget, J., 1975, *L'équilibration des structures cognitives. Problème central du développement*, P.U.F., Paris.

Pinson, G.; Demailly, A.; Favre, D., 1985, *La Pensée. Approche Holographique*, P.U.L., Lyon.

Pribram, K.H., 1977, Holonomy and Structure in the Organization of Perception, in *Images, Perception and Knowledge*, J.M. Nicholas (ed.), Reidel, Dordrecht.

Sauvage, M., 1973, *Parmenide*, Seghers, Paris.

Simon, H., 1986, in *Sciences de l'Intelligence, Sciences de l'Artificiel*, J. Demailly and J.L. Lemoigne (eds.), P.U.L., Lyon.

Stengers, I., 1985, in Débat (4ème Séance, F.J. Varela: L'Auto-organisation: de l'apparence au mécanisme; P. Livet: Fascination de l'auto-organisation), *L'Auto-Organisation*, J.P. Dupuy and J. Dumouchel (eds.), Seuil, Paris, p. 173.

Tabary, J.C., 1986, Sens de l'ouverture, *11ème Congrès International de Cybernétique*, Association Internationale de Cybernétique, Namur, pp. 323-327.

Tabary, J.C., 1987, Le paradigme holographique, *Revue Internationale de Systémique*, vol. 1, pp. 159-180.

Vallée, R., 1984, Eigen-elements for observing and interacting subjects, *Cybernetics and Systems Research*, vol. 22, North Holland, Amsterdam, p. 89.

Varela, F.J., 1987, Experimental epistemology: background and future, *Revue Internationale de Systémique*, vol. 1, pp. 209-224.

Vullierme, J.L., 1987, *Le Concept de Système Politique*, P.U.F., Paris.

Lama Thubten Yéshé and Lama Zopa Rimpoché, 1985, *L'Energie de la Sagesse*, Institut Vajra Yogini.

NON-RATIONAL COGNITIVE PROCESSES AS CHANGES OF DISTINCTIONS*

Francis Heylighen

PESP
Free University of Brussels (VUB)

Abstract

Rational cognitive processes are defined as processes constrained by an external system of rules. This constraint is represented by the conservation of distinctions, where a distinction is conceived as an element of cognitive structuration. Four classes of distinctions (patterns, states, rules and values) and four classes of distinction processes (conservation, destruction, creation and creation-and-destruction of distinctions) are defined. The resulting 4 x 4 grid is used to classify cognitive processes. This allows the modelling of 'non-rational' phenomena, such as creativity, emotions, mystical experiences, etc., in a relatively simple way, as incompletely distinction conserving processes.

1. Introduction

Originally, the concept of 'cognition' was associated primarily with 'rational' forms of mental activity: reasoning, thinking, conscious manipulation of knowledge, deduction, computation, etc. This can be traced back to traditional philosophy, which divided consciousness into three principal functions: cognition, conation and affection. Conation and affection stand here for the mental processes associated, respectively, with volition (motivation, purposiveness ...) and emotion (subjective experience). The first formal or computer models of cognitive processes were indeed restricted to typically deductive or computational ways of manipulating information, and it is still unclear how to model emotional phenomena with a computer.

However, we have come to a point were emotion and motivation are no longer considered to be outside the scope of cognitive science. In psychology, numerous experiments are designed to observe how cognitive structures determine our emotional reactions or motivations (see e.g. Hamilton, 1983; Guidano & Liotti, 1983; Schachter & Singer, 1962). In artificial intelligence, we have moved from

G. Van de Vijver (ed.), New Perspectives on Cybernetics, 77–94.

deterministic algorithms to models of general heuristic knowledge about problem-solving, then to models of the 'intuitive' knowledge used by an expert in his particular problem domain, and we are now beginning to understand how 'activation' or 'arousal' can spread through a network of 'associations' so as to allow for cognition (Rumelhart & McClelland, 1986).

In this evolution, we clearly see a movement from modelling explicit, controlled, deterministic processes ('reasoning', 'deduction', 'computation') towards modelling implicit, spontaneous, indeterministic processes ('intuition', 'association', 'experience'). I have chosen to call the former processes *rational,* the latter *non-rational.* (In a different context I have called the cognitive structures or representations associated with these processes 'classical', respectively 'non-classical', cf. Heylighen, 1990). Since I have not found any unambiguous definition of the term 'rationality', I suspect that some people would object to the use of the word 'non-rational' in this context. Therefore I will propose a general criterion for distinguishing rational from non-rational processes.

A first remark to be made is that such a distinction cannot be absolute or discontinuous. There clearly is some *degree* of rationality involved: one of the first lessons of cognitive science is that no realistic cognitive process can be considered to be either perfectly rational, or completely irrational. This principle is exemplified by Herbert Simon's (1957) concept of *bounded rationality,* which sees human decision-making or problem-solving as a trial-and-error search for satisfying solutions, instead of a systematical examination of all rationally conceivable possibilities in order to find an optimal solution. On the other hand, the recent developments in expert systems (Jackson, 1986) and knowledge engineering have shown that decisions which appear to be made purely by 'intuition', without any rational explanation or argumentation, often can be modelled by systems of 'if ... then ...' rules with a more or less logical structure.

At this point, a critic might argue that there is no real distinction between rational and non-rational cognition. However, it seems 'intuitively' evident that there is a qualitative difference between the mental activities of an engineer who calculates the amount of concrete he needs in order to build a bridge, and a person who is dreaming or freely associating. But how then can we 'rationally' characterize this difference ?

In the first case, the engineer is clearly using a method, a system of rules, which is independent of the specific situation he is approaching. We can say that the process is *externally controlled*: the rules of thought which govern the process are universally or globally applicable to a certain problem class; they are not intrinsic to the situation at hand. At the limit of perfect rationality, this would entail that there be a system of rules applicable to all problem situations (cf. the

Cartesian 'Method', which is at the base of rationalist philosophy). Moreover, such a method would completely determine the process: no room for guessing or doubting would be left. The algorithm the engineer uses for calculating is deterministic: the process can have only one possible solution or result.

In the second case, on the other hand, there clearly is no external control. Seen from the outside, the dreaming process is completely indeterministic or unpredictable: the same initial situation can lead to an infinity of different results. However, this does not mean that control is completely absent. Within a dreaming sequence, the dreamer has the subjective feeling that all transitions between situations are natural or logical. It is only after the experience, when he tries to rationally reconstruct the sequence of events, that he is struck by the seeming absence of any logic or causality. This feature might be described as internal or *local control*: the choice between different possible transitions of cognitive states is made locally, at the moment the particular state or situation appears. The transition from dream situation **A** to the subsequent situation **B** is smooth: in general there is a direct 'association' between **A** and **B**. The same holds true for the transition from **B** to **C**. However, there is no direct relation between **A** and **C**: in general you cannot find any rule or association which would lead from **A** to **C**. If the sequence of situations is considered as a whole, it is impossible to find any global rules governing its dynamics.

The problem we must face then is the following: how can we formally describe this feature of 'local control', so that we can build a framework for representing and classifying the principal 'non-rational' processes of the mind ?

2. *Rationality as Distinction Conservation*

If we wish to model the 'external system of rules' which determines rational processes, we may find inspiration by looking at classical physics, which provides a mathematical theory of deterministic processes. The encompassing rules we are looking for are here called 'Laws of Nature' or 'Fundamental Physical Principles'. The simplest type of laws are called *conservation principles* (Heylighen, 1987). They state that during the evolution of a system, certain global characteristics of this system (e.g. its mass, or its energy) must remain invariant. The cognitive analogue of such a physical principle would state that there is a certain feature of cognitive states which is conserved during all rational processes. However, it seems difficult to find an analogue for measurable physical quantities, such as mass, in the realm of internal, mental processes.

Let us then turn to a second type of laws, the so-called *variation or optimization principles* (Heylighen, 1987). They require that possible trajectories of an evolving system be compared and evaluated according to some overall criterion (e.g. the 'action' associated with the path in phase space). The trajectory to be realized is that for which the value is 'optimal' (i.e. maximal or minimal, according to the definition of the criterion). This is clearly similar to the 'hill-climbing' paradigm used to model the cognitive process of problem-solving: the problem-solver is supposed to choose that operator (or problem state) which is optimal, according to a heuristic criterion, which measures the 'closeness to the goal' of the resulting problem state. The difficulty here, however, is that there does not seem to be a universal criterion, like the 'action' in classical mechanics, which would be applicable to all situations a cognitive system may be confronted with.

Let us move this analysis of dynamical principles to a higher level of abstraction. Both conservation and variation principles are ways of expressing the requirement of *causality*, which may be defined as follows:

equal causes have equal effects

In general we can interpret this requirement to be valid in the two directions, i.e. we may infer that 'equal effects have equal causes'. Taking the contraposition of this latter proposition leads to the following propositions:

distinct causes have distinct effects,

which is an equivalent way of expressing the causality principle. Representing a cause by c_i and its corresponding effect by e_i, this can be expressed more formally as:

$c_1 \neq c_2$ *if and only if* $e_1 \neq e_2$.

This rule can be interpreted as a more abstract conservation principle: the 'distinction' between two initial states (causes) is conserved during the process, and can be found back in the final states (effects). The usual conservation principles (e.g. energy conservation) can be reduced to special cases of this principle: consider two initial states with distinct energies $\mathbf{E}_1 \neq \mathbf{E}_2$, then after a dynamical evolution they will still have the same distinct energies $\mathbf{E}_1 \neq \mathbf{E}_2$ (if $\mathbf{E}_1$ would have become equal to $\mathbf{E}_2$, then energy would not have been conserved). Variation principles, on the other hand, presuppose that it is possible to distinguish 'optimal' (or 'better') state transitions from 'suboptimal' (or 'worse') state transitions, and that this distinction remains invariant during the dynamical evolution. They can hence also be seen as special cases of this principle of *distinction conservation*.

We now have moved to a level of description which is abstract enough to allow a generalization from *physical principles to cognitive principles*. Indeed, 'distinction' is a concept which, unlike 'energy' or 'action', does not require the measurement of concrete physical properties in order to be determined. We will now analyze the general meaning of this concept in the context of cognition. Then we will move back to a more concrete level, and try to find more specific types of cognitive distinctions, which could play roles analogous to the 'energy', 'mass' or 'action' distinctions in physics.

3. A hierarchy of cognitive distinctions

Cognition might be characterized as a structuring of experience or perception, so as to provide structured information ('knowledge') which can be manipulated or processed in order to solve problems. This perceptual structuring can be conceived as a *classification* of phenomena or stimuli. For example, different spectra of light experienced by the optical nerves can be put together in a global class, designated by the colour 'red', which is distinguished from other classes of spectra designated as 'blue', 'green', 'yellow', etc. The fundamental process here is that you have a very large, unstructured set of physically different phenomena (e.g. spectra), which is reduced to a much smaller set of classes (e.g. colours), such that the differences between phenomena in the same class are considered to be negligible or irrelevant, whereas the differences between phenomena belonging to different classes are considered to be important or meaningful. In the first case we can say that the phenomena are 'assimilated', they are considered to be equivalent with respect to the cognitive system. In the second case the phenomena are 'distinguished'.

A *distinction* can be defined as the process (or its result) of discriminating between a class of phenomena and the complement of that class (i.e. all the phenomena which do not fit into the class) (Spencer-Brown, 1969). As such, a distinction structures the universe of all experienced phenomena in two parts. Such a part which is distinguished from its complement or background will hereafter be called an *indication* (Spencer-Brown, 1969). If more than one distinction is applied the structure becomes more complex, and the number of potential indications increases, depending on the number of distinctions and the way they are interrelated. A distinction (which is always implicitly accompanied by the assimilation of phenomena corresponding to the same indication) can hence be seen as an *element of cognitive structuration.*

In order to make a more fine-grained analysis we will now consider four different types of distinctions: pattern distinctions, state distinctions, rule distinctions and value distinctions. The first type is the result of the perceptual process which may be called 'pattern recognition'. This process abstracts away certain minor perceptual differences between stimuli (e.g. the exact position of a spot of light on the retina) in order to find a more meaningful, invariant, common feature (e.g. the global form: square, triangle, circle, etc., of a set of light spots).

In a more formal framework (e.g. the knowledge representation schemes used in artificial intelligence, cf. Jackson, 1986) such a pattern can be represented as a 'structured object' consisting of a functor or predicate **F**, with a list of arguments a_i, which are either atomic objects (variables or constants), or structured objects themselves;

$$\mathbf{F}\ (\mathbf{a}_1, \mathbf{a}_2, ..., \mathbf{a}_n)$$

The use of variable arguments (which can have several 'ground' instances) allows us to represent a class of different objects by one 'pattern' or expression, this pattern is distinguished from other patterns by its structure (number of arguments, number of levels, ...) and by its constant components.

Consider an agent cognitively interacting with an environment; This means that in the present situation, the agent will perceive a certain number of patterns. These patterns may be said to be *actual, indicated* or *activated* with respect to the particular situation the agent is in. The other patterns, which are not activated in the present situation, although the agent is capable of distinguishing them, may be called *potential.* The whole of actual patterns at a certain instant in time (which may be activated by external stimuli, or by the internal dynamics of the cognitive system) determine the *cognitive state* of the agent. Such states determine a new type of distinction: two cognitive states $\mathbf{S}_1$ and $\mathbf{S}_2$ are distinct if there is one (or more) pattern(s) **P**, so that **P** is actual in the state $\mathbf{S}_1$ and potential in the state $\mathbf{S}_2$.

Remark that this distinction is determined by the distribution of activation and not by the 'structure' of the state. In a logical framework, a state may be described as a conjunction of primitive propositions, where such a proposition corresponds either to a structured term **P** (with instantiated variables) representing an actual pattern, or to the negation **P'** of such a term, representing a pattern which is not actual. For example:

$$S = P_1 \ \& \ P_2' \ \& \ P_3' \ \& \ \dots \ \& \ P_n.$$

All different states will hence have the same structure: they will consist of the same list of n terms $P_1, \dots, P_n$. The only difference resides in the distribution of the negation signs ('), indicating absence of activation, over the different terms.

The third type of cognitive distinction corresponds to 'rules' or 'operators'. In order to adapt to the environment, a cognitive system should be able to internally represent possible evolutions of the environmental (or internal) situation. This type of distinction is dynamic: it represents a *change* of the situation or state, it distinguishes previous situations from subsequent situations. As such it has a temporal dimension. Such a change can be described by an operator or function which sends initial cognitive states onto subsequent states.

$$f : S_1 \rightarrow f(S_1) = S_2.$$

More generally, it can be represented as a 'production rule', which takes a conjunction of patterns in as arguments ('conditions'), and produces a new conjunction of patterns ('actions'):

if	$P_1 \ \& \ \dots \ \& \ P_n$,
then	$Q_1 \ \& \ \dots \ \& \ Q_m$.

Remark that in the present, 'logical' representation, the structure of a rule distinction is more complicated than that of a state or pattern distinction. This is not necessary, however; it is possible to conceive representation systems in which processes are more fundamental than states or objects (cf. Heylighen, 1984).

A last type of distinctions is needed to guide the actions of the cognitive agent. It is useless to represent possible changes of the situation, and hence possible actions of the agent, if there is no way to distinguish positive changes from negative ones. From the viewpoint of evolutionary epistemology (Campbell, 1974), the ultimate criterion for evaluating changes is survival: does the conceived change endanger or secure the long-term survival of the agent ? (Heylighen, 1988). In practice, however, the agent needs more local evaluation criteria ('vicarious selectors', Campbell, 1974), which allow him to directly choose between alternatives for action. Such criteria will be called *value distinctions*.

In problem-solving they correspond to the distinction between goal and non-goal states or between 'better' and 'worse' state transitions, as determined by a heuristic criterion. In the General Problem Solver (GPS, Newell & Simon, 1972) these distinctions are called 'differences'. The method of 'means-ends analysis', which is at the base of GPS, consists in establishing a correspondence between distinctions at the value level, and distinctions at the rule level, by looking for a set of rules ('means') which would be able to eliminate the 'differences' which exist between the present state and the goal state ('end').

We now have defined a sequence of four basic types of cognitive distinctions: pattern distinctions, state distinctions, rule distinctions and value distinctions. The order or hierarchy of this sequence corresponds to a 'degree of subjectivity'. It goes from the most 'objective', environment-centred distinctions (patterns) to the most 'subjective', agent-centred distinctions (values) (cf. Heylighen, 1988). We will now develop a classification of processes according to their 'degree of distinction conservation'. By combining the two classification dimensions, we may then develop a scheme for classifying different rational and non-rational cognitive processes.

4. A hierarchy of processes

Rational processes were defined as cognitive processes globally constrained by a system of rules. The effect of this constraint can be conceived as a conservation of the distinctions characterizing the process. However, as remarked in the introduction, it is not very useful to look for an absolute demarcation between rational, i.e. distinction-conserving processes, and non-rational, i.e. non-distinction conserving processes. We would prefer a more fine-grained classification, in which different degrees of 'rationality' could be distinguished. Therefore we will analyze the signification of different types of (incomplete) distinction conservation.

The first type of processes, characterized by a complete conservation of distinctions, corresponds to the processes described by classical mechanics. They are characterized by features such as: causality, determinism, reversibility, absoluteness of space and time, etc. (Heylighen, 1990). This entails that, given the state of the process at a certain instant in time and given the set of rules, all future states of the process can be predicted, and all past states can be reconstructed. It can further be shown that cognitive structures (theories, representations, and so on) incorporating this requirement possess all properties associated with perfect or 'classical' rationality: absoluteness of truth, Boolean logic, optimizing decision-making, 'objective' observation, etc. (Heylighen, 1990). However, there is clearly no realistic cognitive process which obeys all these requirements. In a very rough approximation, by neglecting all trial-and-error steps and by assuming a perfect memory for all necessary information, the computation of the trajectory of a planet by an astronomer, using classical mechanics, might be considered as an example of such a deterministic cognitive process.

A distinction conserving process can be described mathematically by a bijection, i.e. a function f which is *one-to-one*, which maps distinct elements onto distinct elements:

$f(s_1) \neq f(s_2)$ *if and only if* $s_1 \neq s_2$.

In many practical situations, a physicist will shift from classical mechanics to statistical mechanics or thermodynamics. In a thermodynamical situation which is 'near to equilibrium' (Prigogine, 1979), the physicist can still make deterministic predictions about the process. However, the process is in general not longer *reversible*. This means that distinct initial states can lead to the same final state (the equilibrium state, or state with maximal entropy). This is called *equifinality*. Given this final state, it is impossible to determine which of the distinct initial states originally led to this result, hence the feature of *irreversibility:* without extra information it is impossible to reconstruct the initial state, i.e. to reverse the process. The process has 'forgotten' its initial state; there is a loss of information. Such a process is clearly non-distinction conserving: distinctions between states are 'destroyed'. However, it remains deterministic, since there is only one possible final state for a given initial state.

Mathematically such a process can be represented by a surjection, i.e. a function which is *many-to-one*, which can map distinct elements onto the same element, although distinct final elements can only be the result of a mapping of distinct initial elements:

$f(s_1) \neq f(s_2)$ *only if* $s_1 \neq s_2$.

A more complex description is needed for thermodynamical situations which are 'far from equilibrium' (Prigogine, 1979). Here we may encounter the phenomenon of *bifurcation:* the path leading from initial states to final states bifurcates or branches. This means that the same initial state can lead to distinct final states. Hence the process is no longer predictable: we do not know which of the alternative paths the process will choose at the bifurcation point. Such a process is characterized by the 'creation' of distinctions, and hence of information. This aspect of creativity is illustrated by the appearance of dissipative structures in far-from-equilibrium thermodynamical systems.

If we consider a distinction-creating process without distinction destruction (which is not realistic in thermodynamics), then it can be represented by the inverse of a surjection, i.e. a relation (it is no longer a function), which is *one-to-many*. This means that the same initial element can be sent to distinct final elements, although distinct initial elements will still be sent to distinct final elements:

$$\mathbf{f}(s_1) \neq \mathbf{f}(s_2) \textit{ if } s_1 \neq s_2.$$

We finally come to the general case, where there is both creation *and* destruction of distinctions. Such a process is neither predictable nor reversible. There is both a creation of new information and a forgetting of old information. Such processes may be called *self-organizing:* there is no external system of rules controlling the process, its development is determined locally or internally, by the evolving system itself (Heylighen, 1988). They can be represented mathematically by *many-to-many* relations, this time, however, without any global rule expressing a correspondence between initial and final distinctions.

Let us summarize this section.

We have introduced four basic types of processes, characterized respectively by: (I) distinction conservation (one-to-one), (II) distinction destruction (many-to-one), (III) distinction creation (one-to-many), and finally (IV) distinction creation-and-destruction (many-to-many). The order of the sequence can be seen as indicating a 'decrease in rationality'.

Processes of the first type are completely controllable: they can be both predicted and retrodicted or reversed. Cognitive instances of such processes would correspond to logical or causal inferences, or algorithmic computations, in which there is a complete conservation of information.

The second type of processes can only be controlled in the 'forward direction': they are predictable, but not reversible. In the cognitive domain they would correspond to inferences which are still deterministic, but where there is a loss of information (e.g. abstraction, forgetting, habituation, etc.).

The third category is controllable in the backward direction: the processes are reversible, but cannot be anticipated. A cognitive process of this type is characterized by 'creativity', by a step of which the result is unpredictable, but which may lead to new knowledge. This knowledge, together with the already existing knowledge, is maintained in memory: the process is accumulative, no information is lost.

For the final category, there is not any global control in either direction: the process is unpredictable and irreversible. Cognitive processes of this type are characterized by both a creation of new knowledge and an elimination ('forgetting') of old knowledge.

5. A classification scheme for cognitive processes

The classification of processes we have made is still too rough to be very useful for describing cognition. As we have said, complete distinction conservation is unrealistic. But also distinction conservation in one direction (only creation, or only destruction of distinctions) is not very realistic. This leaves us with the last category of processes. The problem here is that there is not any dynamical principle left which could help us to describe the process by constraining the set of potential models: anything is possible.

Scientific models are by definition rational: they attempt to propose rules for explaining and predicting phenomena, which are explicit and universal, i.e. which are applicable not to one particular phenomenon, but to a class of similar phenomena, which is as large as possible. Hence a scientific theory corresponds to what we have called an 'external system of rules', characterized by 'distinction conservation'. We are then faced with a paradox: how can we model non-rational processes in a rational way ?

This paradox may be resolved by noting that rationality is not an absolute or universal feature, but a local one: the same cognitive process may be rational in certain aspects and non-rational in other aspects. A useful rational theory may then provide a representation of all rational aspects of the process, and of the relation or interaction of these aspects with the other ones. In this way, the theory of the phenomenon is as complete as possible, without becoming irrational itself. In our distinction-based framework, this may be realized by studying which distinctions are conserved by the process, and in which direction (forward or backward or both). This should allow us to build models which are partial, but rational.

This principle may again be illustrated by an analogy with physics. In most practical physical phenomena, there is friction (i.e. dissipation of energy). This means that energy is not conserved, and hence the reversible laws of classical mechanics are not applicable. Yet in many cases we may assume that other quantities characterizing the system, such as mass or charge, are conserved. Hence we can still make certain predictions, although the model is not completely classical. In the same way, one can consider thermodynamical systems which are 'open', such that neither energy nor mass nor charge are conserved. Yet they may be characterized by invariant 'boundary conditions' which (partially) determine their dynamics.

The approach to follow in such cases is to *factorize* the description of the system into a set of properties or distinctions, such that the invariant distinctions are separated from the variable ones, and such that the subset of (independently) invariant distinctions is as large as possible. This set of stable distinctions can then be used as the basis for a dynamical model which is rational.

This is what we have attempted to do, when we defined four types of cognitive distinctions. This is of course still a first approximation: more detailed theories of cognition will require more elaborate classifications of distinctions, and perhaps of distinction dynamics. However, we will now show that these four classes, together with the four classes of processes, are sufficient to construct a frame in which some of the most interesting cognitive phenomena, including difficult to model because 'irrational' phenomena, can be situated and identified.

The Cartesian product of the two sets of four elements produces a matrix containing sixteen elements (see Fig. 1). A cognitive process can now be characterized by indicating the corresponding square(s) in this 4 x 4 grid.

	conservation I	destruction II	creation III	creation and destruction IV
Pattern distinctions	I.Pa	II.Pa	III.Pa	IV.Pa
State distinctions	I.St	II.St	III.St	IV.St
Rule distinctions	I.Ru	II.Ru	III.Ru	IV.Ru
Value distinctions	I.Va	II.Va	III.Va	IV.Va

Figure 1

A 4 x 4 classification scheme for cognitive processes, with distinction types as the vertical dimension and process types as the horizontal dimension.

For example, a process which conserves state distinctions is represented by the square I.St. A process creating rule distinctions corresponds to the square III.Ru. Of course, many processes will be characterized by more than one square. In general, for each of the four distinction types, Pa, St, Ru and Va, there will be a corresponding dynamical behaviour: I, II, III, or IV. Hence, a cognitive process may in general be represented by a function from the set of distinction types to the set of process types:

$$\mathbf{f} : \{\mathbf{Pa},\mathbf{St},\mathbf{Ru},\mathbf{Va}\} \rightarrow \{\mathbf{I},\mathbf{II},\mathbf{III},\mathbf{IV}\}.$$

The set of all such functions, however, is too large to be discussed element by element (its cardinal number is $4^4 = 256$). Therefore, we will discuss the grid more roughly, square by square, giving examples of processes for the most interesting squares.

6. A discussion of some non-rational process classes

A process for which all squares are in the first column (I) of the matrix is perfectly rational. As we have discussed, this is not very realistic. However, many useful models of cognitive processes assume that all squares are in the first column, except the one corresponding to the state distinctions (St). Destruction of state distinction (II.St) is a model for 'goal-directed' processes, where the 'goal' or 'end' of the process consists in reaching one (or a few) final state(s). This is typical for biological organisms, characterized by homeostasis. The goal itself is hereby determined by the (conserved) value distinctions. Creation of state distinctions (III.St) is a model for processes characterized by 'trials', by exploration of possibilities without deterministic guidance. The combination of both mechanisms (IV.St) is typical for problem-solving models, based on a trial-and-error or generate-and-test principle. The trial phase corresponds to a creation of distinctions, the error or test phase corresponds to a destruction of distinctions (Heylighen, 1988).

These models are still 'relatively' rational. Indeed, states are by definition more variable than the other cognitive structures (patterns, rules and values), so we are not surprised that distinctions between states would change. The formalization of such models is relatively simple, and resembles the formal structure of thermodynamical models. In computational terms, states correspond to 'data', whereas rules, patterns, and values correspond to 'programmes', which control the processing of data. It is clear that it is more difficult to model processes which change their own programmes. In psychological terms, states belong to the 'short-term memory', whereas the other distinctions belong to the 'long-term memory'.

Let us consider a change in pattern distinctions. The destruction of a pattern distinction (II.Pa) corresponds to the process which is called 'habituation' in learning psychology (Wickelgren, 1977). This means that the cognitive system learns to ignore (i.e. not to distinguish) a particular stimulus pattern. The opposite phenomenon (III.Pa) is called 'sensitization': the system is sensitized to distinguish a new pattern. In higher level cognitive systems this phenomenon is called 'concept learning': the subject discovers a new concept or category allowing it to structure its perception. The combination of both phenomena (IV.Pa) is typical for processes of cognitive development or discovery.

Complementary to pattern learning there is rule learning (III.Ru). A rule can be conceived as an association between distinct patterns. It is only meaningful to learn a new pattern if this pattern is associated to other patterns. Of course, rules can also be 'unlearned' (II.Ru).

The learning of a value (III.Va) could be conceived as the learning of a pattern which is directly associated with an existing value distinction or, on a more primitive level, with the physiological system which distinguishes pain from pleasure. A value can be unlearned (II.Va) if this association is weakened. The long-term process which is called 'personal growth' consists of a creation of new values together with a rejection of old values (IV.Va) until a more or less stable, well-adapted value system emerges. In psychotherapy the rejection of inadequate value distinctions is sometimes called 'transcendence of dichotomies'. Consider for example the distinction (or dichotomy) between being 'brave' and being 'prudent'. Both sides (indications) of the distinction may appear valuable. However, they seem to contradict each other. A person with a psychological problem may then hesitate continuously between the two alternative types of behaviour: should he act bravely or prudently ? The solution of the problem consists in transcending this absolute distinction and replacing it by one or more distinctions which are better adapted to the situation: e.g. one should act bravely in situations where he can anticipate what will happen, and act prudently in situations where everything is uncertain.

We should further say something about processes in which more than one type of distinction is changed. These are the processes which are the 'least rational'. General, long-term processes such as 'creativity' or 'cognitive development' belong to this category. But also short-term phenomena like emotions or 'mystical' experiences can be modelled in this way.

Cognitive theories of *emotion* (Schachter & Singer, 1962) posit two components of an emotional experience: (physiological) arousal, and a cognitive interpretation of the situation causing this arousal. In a more detailed model (Davitz, 1969) four components are distinguished: activation or arousal, hedonic tone (pleasure or displeasure), relatedness, and competence. The two last components are cognitive interpretations or classifications of the relation between the subject and his situation, and might be modelled in our framework by specific, invariant patterns (I.Pa). Arousal (at least its purely mental component) can be seen as enhanced search activity or 'mental exploration' (III.St). Pleasure or displeasure can be seen as enhanced 'goal-directedness' (II.St). What is typical of an emotional process is not that state distinctions are not conserved (this also happens during general problem-solving), but that there is a clear switch from the normal mode of change of state distinctions to another one, characterized by more (or sometimes less) arousal, and by different goal-directed behaviour. This could be modelled by assuming that new value distinctions are activated by the particular situation which causes the emotional response. Hence the overall process may be characterized by the squares IV.St and IV.Va.

A so-called *mystical experience* (cf. Deikman, 1966; Tart, 1969), which may be provoked by meditation or by the intake of hallucinogenic drugs, is characterized by a drastical release of very basic distinctions, e.g. the self-world distinction and the past-future distinction. This is experienced as a feeling of wholeness or unity between the self and the cosmos (the so-called oceanic feeling). This corresponds to a (temporary) destruction of patterns (II.Pa). However, without patterns, the corresponding rules and values must also disappear from consciousness, hence we may add the squares II.Ru and II.Va: the meditator looses any sense of goal-directed action. Such a radical, distinctionless mode of consciousness can, however, not be maintained permanently. The typical reaction, after the experience, is that familiar phenomena are perceived in a new, fresh way; the world is seen differently, perhaps with more detail, with more vividness ... (Deikman, 1966). This can be explained by assuming that in its normal, equilibrium state the mind must maintain a certain amount of distinctions. If some of the usual pattern distinctions are suppressed through special circumstances (e.g. prolonged concentration on one specific phenomenon), the mental system will tend to counterbalance this 'alteration of the equilibrium' by making new distinctions. Of course the mind may make the same distinctions as before the experience, but one of the aims of prolonged meditation is just to *de-automatize* this process, so that distinctions different from the usual ones may get a chance to become actualized (cf. Deikman, 1966). If we wish to include this effect in our scheme, we must move to column IV.

There finally is a very large, undifferentiated class of non-rational cognitive processes which may be labeled as 'intuitive', 'associative', 'experiential' (cf. Gendlin, 1962), 'subconscious', etc. Dreams or phantasies are examples of such processes. These are processes for which there is no general class of distinctions that are conserved. Everything seems to be fleeting, transient.

Consider for example a dream in which you are talking to a person. The identity of the person is not clear, however. At the beginning of the sequence you had the impression that you were talking to a former girl friend. By the end of the discussion you are definitely conversing with your sister. Yet you cannot remember any point in the dream sequence where your girl friend changed into your sister. There is not any rule distinction which could represent such a change. Obviously the distinction between girl friend and sister was not conserved during the dreaming.

This does not mean that such phenomena cannot be modelled in a rational way, however, or that there are no distinctions determining the process. The problem here is that the stability of distinctions is very local or limited. Specific distinctions are conserved by specific processes during specific time intervals, but changed during different intervals or by different processes. We need a much more detailed scheme for modelling such processes. Two approaches seem to be promising. As we remarked earlier, such locally controlled processes are still continuous. Continuity can be defined mathematically as the conservation of 'topological' distinctions (Heylighen, 1990). This should inspire us to study the topology of cognitive structures or semantic fields, so that we could introduce new distinction types. A second approach would consist in building a more detailed classification of processes, which would propose more classes besides bijections, (inverse) surjections, and general relations. This last category indeed includes the most diverse network structures, which may be useful for modelling many types of locally controlled processes (e.g. through the paradigms of connectionism or parallel distributed processing, Rumelhart & McClelland, 1986).

7. Conclusion

We have argued that cognitive processes can be characterized by the distinctions (i.e. elements of cognitive structuration) they make, and by the way such distinctions are conserved. A rational cognitive process was defined as a process constrained by an external system of rules. This feature of global constraint could be represented by requiring that all distinctions be conserved during the process. In this sense, however, real cognitive processes can only be *partially* rational: they only conserve certain distinctions, perhaps only in one direction (forwards or backwards), and perhaps only during a limited time interval. This feature of incomplete rationality instigated us to undertake the construction of a more detailed classification of cognitive processes, by asking which type of distinctions is conserved by which process in which way. The more distinctions a process would conserve, the closer it would come to the rational ideal; the less distinctions it would conserve, the more it would deserve the label of 'irrationality'.

In order to do this we needed to distinguish different classes of distinctions and classes of distinction processes. We introduced four distinction types: patterns, states, rules, values, and four process types: conservation, destruction, creation, creation-and-destruction. In this way we laid the base for a new rational model of non-rational processes. Indeed, the classes which were defined are (conserved) distinctions themselves, but distinctions about distinctions, i.e. *meta-distinctions*. Hence some form of rational representation of non-rational phenomena may be attained by moving to a meta-representational level.

The further elaboration of these models can be conceived as the construction of a *meta-representation* (Heylighen, 1988, 1990). Such a theory could be formalized and operationalized through a computer implementation and a psychological experiment (Heylighen, 1988).

* This article was first published in *Communication & Cognition*, vol. 23, nos. 2-3.

References

Campbell, D.T., 1974, Evolutionary Epistemology, in *The Philosophy of Karl Popper*, Schilpp, P.A. (ed.), Open Court Publishing, La Salle, Illinois, p. 413.

Davitz, J.R., 1969, *The Language of Emotion*, Academic Press, London.

Delkman, A.J., 1966, De-automatization and the Mystic Experience, *Psychiatry*, vol. 29, p. 324.

Guidano, V.F.; Liotti, G., 1983, *Cognitive Processes and Emotional Disorders*, Guildford Press, New York.

Gendlin, E.T., 1962, *Experiencing and the Creation of Meaning. A philosophical and psychological approach to the subjective*, The Free Press of Glencoe.

Hamilton, V., 1983, *The Cognitive Structures and Processes of Human Motivation and Personality*, Wiley, London.

Heylighen, F., 1987, Towards a General Framework for Modelling Representation Changes, *Proceedings of the 11th International Congress on Cybernetics*, Association Internationale de Cybernétique, Namur, pp. 136-141.

Heylighen, F., 1988, Formulating the Problem of Problem Formulation, in *Cybernetics and Systems '88*, R. Trappl (ed.), Kluwer, Dordrecht, pp. 949-957.

Heylighen, F., 1990, *Representation and Change. A Metarepresentational Framework for the Foundations of Physical and Cognitive Science*, Communication & Cognition, Ghent.

Jackson, P., 1986, *Introduction to Expert Systems*, Addison-Wesley, Wokingham, England.

Newell, A.; Simon, H.A., 1972, *Human Problem Solving*, Prentice-Hall, Englewood Cliffs.

Prigogine, I., 1979, *From Being to Becoming: Time and Complexity in the Natural Sciences*, Freeman, San Francisco.

Rumelhart, D.E.; McClelland, J.L. (eds.), 1986, *Parallel Distributed Processing: Explorations in the Microstructure of Cognition*, Bradford Books/MIT Press, Cambridge, MA.

Schachter, S.; Singer, J., 1962, Cognitive, Social and Physiological Determinants of Emotional State, *Psychological Review* , vol. 69, p. 379.

Simon, H.A., 1957, *Models of Man: Social and Rational*, Wiley, London.

Spencer Brown, G., 1969, *Laws of Form*, Allen & Unwin, London.

Tart, C.T. (ed.), 1969, *Altered States of Consciousness*, Wiley, New York.

Wickelgren, W.A., 1977, *Learning and Memory*, Prentice-Hall, Englewood Cliffs, N.J.

EPISTEMOLOGICAL ISSUES

SELF-ORGANIZATION AND AUTONOMY IN A POST-CYBERNETIC PERSPECTIVE EPISTEMOLOGICAL ISSUES

Fernand Vandamme

Labo for Applied Epistemology - BIKIT
University of Ghent

Abstract

Knowledge of a domain can be seen as a structuration of a domain. Maximal efficient knowledge of a domain can be seen as the minimal structuration which is sufficient for fulfilling the relevant functions, taking into account the context and situation in which the domain is embedded. This minimal structuration guarantees a maximal adaptivity to new opportunities or threats of the environment, without falling into chaos. This requirement of minimality can eventually be loosened locally for the sake of protection of the 'self', self being defined in terms of a symbolic identification of the unity of the structure. The structure can be a culture, a political entity, a research group or an individual. Relative to the optimal minimal structuration, dependent on and relative to an environment, we can define in this context self-organizational options, in terms of operators that eliminate structures, generate structures, or a combination of both.

1. Introduction

With Pask, we agree that the old cybernetics, the new cybernetics and, we like to add, the (old and new) cognitive paradigms are not that revolutionary different from each other. And, as is mostly the case, the so-called new paradigms are in a sense 'older' than the 'old' paradigms. The so-called '*old*' paradigms were in most cases strategically successful specializations in a general framework. Their success was based on a strong but useful simplification of the issues. The '*new*' paradigms are further specializations in the earlier one, or (as is mostly the case) a strategic retreat which broadens the specialized approach and is a return to the original, broader inspiration and outlook. This is what happens, we believe, with the new cybernetics, the post-cybernetics, as well as with the new cognitive approach, and it will presumably also happen with the post-cognitive approach.

G. Van de Vijver (ed.), New Perspectives on Cybernetics, 97–103.

But in this paper we don't intend to discuss labels, although we think that this is relevant and interesting, and not only from a historical point of view. It is, on the contrary, the evolution and development of paradigms, of methodologies or theories, which interest us here: the dynamics of systems, the organization, the self-organization of systems with its internal and external forces in interaction in an ecosphere. Scientists, scientific theories, technological realizations, research groups, private and non-private laboratories, departments, universities, etc., are all systems with a more or less self-organizational behaviour, which are a part of an ecosystem and form together an ecosphere with self-organizational behaviour. It is in this view that we want to have a look at some crucial tools to approach and understand the 'ecosphere' of self-organization.

2. *On Knowledge and Organization*

Knowledge of a domain is the structuration of the domain. If we state this, then we should immediately differentiate between i) '*know-how*' and '*know-that*'[1], ii) knowledge in action and knowledge in vitro, but also between iii) the 'representational' interpretation of 'symbolization' and the 'presentation'/'formative' or action-interpretation of symbolization. It is crucial to be aware of all these differentiations when tackling organizational problems, including the ones of self-organization.

Traditionally, many scientists have a rather short-sighted vision on knowledge. Knowledge is in their view composed of statements about a domain. This is the meta-interpretation of knowledge. Such knowledge corresponds with knowledge of the type 'know-that' and is even, in most cases, knowledge in vitro.[2] This quite easily leads to a dualistic interpretation of knowledge, with a naive correspondence interpretation of truth and a representation vision of meaning. We have, on the one side, the system, the reality, the object (or other such terms), and at the other side we have the description of it: knowledge. We call this view the meta-knowledge approach.

A much richer but also much more performant view on knowledge, is the one which considers the meta-knowledge approach as an important dimension of knowledge, but nevertheless only as a secondary dimension. Knowledge is in the first place a structuration by action, in action. The symbolizations used are not descriptions of objects, structures, but they are these objects, structures, etc. The descriptions create the object, the system. They do not describe the reality, they create, they are the reality. The dualistic approach is here not relevant at all.

Epistemology in this view studies in the first place the construction techniques on this first level.

This view can be found back in the post-cybernetical approach, e.g. in connectionism, but it is also already present in the strict nominalistic approach of Protagoras for instance. The sophists, who laid the first germs of epistemology stressed knowledge in actu. This makes us understand why they rejected the correspondence theory of truth. The efficiency theory of truth is what matters in a knowledge in actu approach. The latter is also what matters in applied epistemology, the former together with coherence is more important in pure epistemology.

Even in a multi-layer approach of knowledge and organization, the differentiation of knowledge in actu and meta-knowledge is relevant and useful! It is also important to be aware of the interaction between both. The difference between knowledge in vitro and knowledge in actu is the heuristical basis for the explicitation of meta-knowledge. But at the same time, meta-knowledge can heuristically be very relevant for the elaboration of the basic structuration of the object domain, of the knowledge in vivo/in actu.

But here again we must warn against an oversimplification that is quite popular but is in fact basically an erroneous approach. The oversimplification is the following: progress is identified with the elaboration of a domain structure: the more a system is structured (be it on the object or meta-level), the better this is. This simple correlation has to be rejected firmly, and this is an important tenet of post-cybernetics, but it is also true for strict nominalism. *Maximal efficient knowledge of a domain, a system, etc., is the minimal structuration that is sufficient for fulfilling the relevant functions, taking into account the context and situation in which the system or the domain are embedded.* This minimal structuration guarantees a maximal adaptivity to new opportunities or threats of the environment.

Let us explain this point. Knowledge in vivo (in actu) is organization, be it eventually supported by meta-knowledge. But organization can be functional or dysfunctional. The functionality can be evaluated in the short term, middle long term, the long term. An organization is also a dynamic process. An organization fulfills a function in a specific situation or context. We can consider an organization as a micro-organization in a macro-organization and so on. Both have a dynamic dimension. The evaluation of the functionality of an organization has to take into account the dynamics, the learning dimension if you want. Learning can be internally or externally motivated. In most cases, it will be with a certain priority of the one over the other. If internally motivated, it will be in terms of the amelioration of a certain internally evaluative function. If externally motivated, it will be in terms of an adaptation to a change of the environment. This change can be a better structuration of it. However, it can also be a qualitative or quantitative change of the external structure.

It will be clear however, that if learning — this means, a change of structure — is of crucial importance for knowledge, that is, for organization, then, in evaluating existing knowledge (i.e. organization), we have to take into account its influence on learning. Does it hamper the dynamics of the learning ? Or does it on the contrary stimulate learning, the realization of the dynamics which is needed. In the latter case, up to the degree that knowledge (structuration) favours *relevant dynamics*, it is beneficial. Up to the degree that knowledge (structuration) hampers *relevant dynamics, knowledge* is obnoxious!

For sure, in practice, the notion of '*relevant dynamics*' has to be specified and elaborated. The treatment of this notion of relevant dynamics can and has to be approached taking into account the specific domains and levels. A general answer and treatment of relevant dynamics is here, as it is the case with knowledge in general, inevitably vague. But anyway, *relevant dynamics* has to be seen and defined in terms of sustaining existence and development of local and global sub-systems of the ecosphere. The notion of *self-organization* is clearly needed here, for it is relative to the 'self', a particular substructure, that relevance is generated.

3. On Self-organization and Knowledge

The 'self' can be considered as an in vivo sub-organization (of knowledge), with a symbolic identification value for the unity of the system. A system has many aspects: sub-systems. All of them are in continuous adaptation, in movement. Inevitably, each system tends to evolve towards chaos if a symbolic identification of the unity of the structure does not exist. *Therefore, for each system, it is vital that there exists such a centre of symbolic unification. What is chosen as such is in the short term irrelevant*, but in the long term it is of importance for survival reasons. The reason for this is rather simple. The kernel of a system, the 'self' — symbolizing the unity of it — will function as a kernel of a much more elaborated structure: organization. This inevitably means that the dynamic processes will be much more complex and subtle here. Therefore important changes will occur here much slower, they will take much more time. As a consequence, what is chosen as the kernel, will have a crucial influence on the specific dynamics the system will realize. This in turn will influence strongly the dynamics of the macro-structure, as it is a part of it.

We want to stress that the system or the macro-system can be a culture, a political entity, a research group, a religious group, an individual, a computer system, a knowledge technology system, etc. In all of these systems the same problems of dynamics arise. The identifying kernel around which the system is organized, is everywhere the pivot in terms of which relevance is generated. The whole discussion of organizational closure but informational openness[3], has to be situated here. Organizational closure means that the operators which produce the organization and the structure are itself produced by the organization. But openness is required as the operators or even the structures can be taken up, got from, eventually motivated by external organizations[4] or systems (super organizations or parallel organizations). This openness can be formulated here in terms of the self. The productive operators do not have to belong to the self: the symbolic identifying subsystems. We could even, in a weaker sense, define the openness needed in an organization as follows: the self may not only make the active operation of the operators impossible, but may also render the introduction of new ones impossible. For sure, the self, as the unifying principle, may control them, but that is another matter. For all these reasons we can agree on interpreting self-organization in terms of *operational closure*, but this requires structural openness on the side of the possible arguments of the operators, as well as on the side of the operators generated and used. On both levels dynamics is possible and needed and herewith we touch the vital domain of self-organization operators, in all their dimensions.

4. *On the Self-organization Operators*

The knowledge organization operators are multiple. We want to be very short in this article on this point, although an elaborated study is certainly justified and needed. In the first place we can mention *structure generating operators*, but we believe that *structure eliminating operators* are as vital. We have to take into account that, relative to the optimal relevant maximal structuration, one can define over-organization and under-organization. In case of over-organization one has to eliminate structure.

Besides the structure generating and the structure eliminating operators, we can also have *combined operators*: operators which combine, alter, generate a new structure by eliminating an existing structure. But all these structure-operators can be active on the formation level of the organization or on the constraint level of the organization. Constraints are very powerful performative mechanisms for guiding structuration inside a domain or system. But inside the constraints, we can also introduce classificatory operators, analogy operators, deduction, induction, abduction, repetition operators, linking operators, etc. The description of the qualities of the operators, the possibility of combining them, the dynamics present in them, their possible variants (from the weakest form up to the strongest form of each) are very important to study in general as well as in situ and in vivo.

Notes

1. We can even further introduce 'know-what', which is on a level in between 'know-that' and 'know-how'. See F. Vandamme 1991.

2. 'In vitro' means the artificial cultivation of living tissue outside a body. With 'knowledge in vitro' we refer to knowledge which is also artificially 'cultivated', knowledge as embedded in the complex structure and far away from the actual knowledge (knowledge in vivo or in actu). Nevertheless, sometimes such knowledge can be successfully embedded and can influence the overall knowledge and knowledge in actu.

3. Maturana is talking here about structurally openness instead of informational openness.

4. An organization is a structure of structures in which a sub-structure has a symbolic (self-identification) function. This sub-structure can change in time and place.

References

Vandamme, F., 1985, *A Primer on Expert Systems*, Ghent, De Sikkel, Communication & Cognition.

Vandamme, F., 1987, *Capita Selecta uit de Toegepaste Epistemologie*, Ghent, Communication & Cognition.

Vandamme, F., 1991, *Elektronisch encyclopedisch woordenboek voor geavanceerde informatica, toegepaste epistemologie en cognitieve wetenschappen*, Ghent, Communication & Cognition.

Van de Vijver, G., 1990, Teleology, error and explanation in the Standard Models of P.D.P., *Communication and Cognition - Artificial Intelligence*, vol. 7, no. 2, pp. 185-205.

Van de Vijver, G. (ed.), 1990, *The old and new in cybernetic fashions, Communication and Cognition*, Vol. 23, nos. 2/3.

Van de Vijver, G., 1991, *Van Cybernetica naar Connectionisme. Een epistemologische studie van doelgerichtheid*, Academia Press, Ghent.

THE EXPERIMENTAL EPISTEMOLOGY OF WALTER S. MCCULLOCH A MINIMALISTIC INTERPRETATION

*Gertrudis Van de Vijver**

Seminar of Logic and Epistemology
University of Ghent

Abstract

When cybernetics entered the scene during the forties, high ambitions immediately arose and quite unusual claims were made as compared to those of traditional epistemology: cybernetics would do away with the distinction between 'mind' and 'body' (Papert, 1965); it would bring a new interpretation, through the artefact, of the Kantian synthetic a priori[1]; cybernetics would give, following Turing's approach of computability, a mechanistic sense to Kantian schematism (Dupuy, 1985, pp. 105-106).

In this text we shall analyse the statute of that sort of claims. In other words, we shall show what is the relation between epistemological questions which we call classical or theoretical and those occurring within cybernetics, and more in particular in the work of McCulloch. McCulloch was clearly an advocate of *experimental epistemology*. He emphasized, in contrast with Kant for instance, that the experimental inquiry into the functioning, the emergence and the consequences of knowledge can be important for epistemology in general. Furthermore, we shall inquire how McCulloch's experimental epistemology may be relevant to classical epistemology.[2]

Firstly, we shall discuss the position of McCulloch within cybernetics. Secondly, we shall deal with the meaning of his net of formal neurons and with the meaning of the heterarchic nets. This will allow finally us to illustrate in what sense McCulloch is to be situated outside first cybernetics and how he anticipates second order cybernetics. It will also allow us to show that the experimental epistemology which he defends can get an interpretation beyond the frame of a traditional reductionism. What we are proposing here is a minimalistic interpretation: the epistemology, insofar as it tends to be experimental, doesn't have to serve as a support for reductionism, but rather aims at pointing out concrete limits within which a theoretical epistemology can be developed.

G. Van de Vijver (ed.), New Perspectives on Cybernetics, 105–123.

1. McCulloch and the first cybernetics

Warren Scott McCulloch can be characterized concisely and quite accurately as the man of the following question: 'What is a number that a man may know it, and a man that he may know a number ?' His entire career may be almost phantasmatically mapped out on the strength of this question.[3] The construction of formal neurons, for which he is widely known, is considered by him to be an answer to this question. He writes in 1961: "... we have come far enough to define a number so realistically that a man may know it, and a man so logically that he may know a number. (...) I have done more, by indicating how we may do it without error, even though his component neurons misbehave." (McCulloch, 1965a, p. 14).

The question of the number is, not only in a theoretical but also in a practical sense, what McCulloch is really concerned about. The 'embodiment' in the title of the collection of articles *Embodiments of Mind*, is directed at describing coherently and relating on all levels of interpretation the complex of events involved in situations in which one may suppose that a 'mind' is at work (Papert, 1965, p. xiii). This concern about the 'embodiment' not only bears witness to the project for which a logical automaton is considered to be an 'incarnation' of a kind of organic thinking-machine in the central nervous system, but also it bears witness to the restriction which McCulloch accepted from the outset. He readily puts aside all non-rigorously describable elements of that nervous system, and in addition he considers his project not so much to be driven by a pragmatic purpose, but by the *genuinely theoretical problem* of the 'number'.

This could give us the impression that McCulloch's interest was in the first place of a theoretical epistemological nature. The 'embodiment' would thereby chiefly refer to the construction of 'idealized' 'possible' networks, much more than to their actual incarnation in a particular brain.[4] McCulloch's position is not that straightforward though. The specific significance of his net of formal neurons as well as his interest for the heterarchic nets show us why this is not so.

2. 'A Logical Calculus of the Ideas Immanent in Nervous Activity'

Descartes had stated already that the combination of matter and movement could give rise to certain specific qualities. With him, as well as with most of his followers (Julien de Lamettrie forming an exception) this implied a dualism; an actual understanding of consciousness on that basis was excluded. Warren McCulloch's stance here is much more radical. The determination and the inertia of matter are in our century no longer that evident, nor is it evident that intelligence and matter should be considered to be incompatible. Earlier, Russell had already remarked in the same line, that the explanation of mind becomes more materialistic as matter itself becomes less material.[5] That the idea of a combination of matter and motion brings us further than one ever had dared to suppose, that we are thereby brought to accept a monism that is indifferent as to the choice between mind or matter, was clear proof, according to McCulloch, for cybernetics as a whole, and in particular for his net of formal neurons.

Since as early as 1923, he was governed by the idea of representing psychic events via a propositional logic that found its starting point in small sub-parts, the 'psychons'.[6] The formal neuron is nothing but the further elaboration of this idea, and in particular with a view to the description of logical connections between electronic components (McCulloch & Pitts, 1943).

In 'A Logical Calculus of the Ideas Immanent in Nervous Activity', in which a net of formal neurons is described, the notation system of Hilbert and Ackermann (1927), and Carnap (1938) is used. This implies not only that the text is not easily accessible, but also that it departs from a two-valued logic, which expresses the fundamental idea that the most important relations in psychology have an all or none character.

What, now, is a formal neuron, in the sense of McCulloch and Pitts, and what is the epistemological significance of a network which consists of formal neurons ?

Every idealized formal neuron is an elementary arithmetical element of computation which computes a Boolean threshold function. It is a 'many input/single output' threshold device, liable to absolute inhibition, and its output is a logical function of the input. A formal neuron can be switched on or off. The on- or off-state of the input represents the truth or falsity of a proposition. The on- or off-state of the output represents the truth-value of the logical function which is computed by the element. As Daniel Andler states in 'From paleo to neo-connectionism': "... a neuron *is* (or *embodies*) a connective, and its activity *represents* the result of applying the connective to the propositions which are fed to it". As it is possible to implement logical connectives, likewise it is possible to implement almost every conceivable logical combination of elementary propositions in a network. For McCulloch therefore, the activation of a neuron in the brain stands for the truth of a proposition about the external reality. Elementary propositions about reality are verified by sensors. The neurons themselves, which are connected with the sensors, can be more complex combinations of these propositions. A net consists of a number of formal neurons, of which the mutual relations are defined on the basis of rules: rules of connection and rules of operation.[7]

The most important theorem that the authors want to prove in 'A Logical Calculus of the Ideas Immanent in Nervous Activity', can be formulated, in accordance with Von Foerster, as follows: '*Any behaviour which can be logically defined or described, unambiguously and in a finite number of words, can be realized in a formal net of McCulloch and Pitts.*[8]

This theorem states that the behaviour of a network can be described in terms of propositional logic, and that, conversely, within certain restrictions, every logical statement can be expressed in a network. The inspiration from the article of Turing, 'On Computable Numbers, with an Application to the Entscheidungsproblem', is quite visible here. Where Turing had shown that a machine could be construed that could compute any number — 'if a number can be computed by an organism, then it is computable by the Turingian definition of computability, and the other way round' — McCulloch and Pitts proved that neurons that can be excited or inhibited, can extract any configuration of signals from the input.[9]

There can be some disagreement though, as Papert remarks, as to the notion of 'description', which Von Foerster used to reformulate the general theorem. When an unambiguous and complete description is nothing but an explicit enumeration, then simple nets (without circularities) will suffice; when on the contrary any mathematical means can be used in the description, then it is possible that a Turing machine is necessary. McCulloch and Pitts know very well that their nets cannot compute all the functions that can be computed by a Turing-machine. One of the difficulties in this regard is the implementation of *infinite* combinations of elementary propositions that can be represented in networks of simple neurons, with feedback. McCulloch and Pitts admit that, in order to have a real equivalence, both a mobile scanner and a potentially infinite tape should be added to the machine. The conclusion of their article is as follows: "It is easily shown: first, that every net, if furnished with a tape, scanners connected to afferents, and suitable efferents to perform the necessary motor-operations, can compute only such numbers as can a Turing machine; second, that each of the latter numbers can be computed by such a net; and that nets with circles can compute, without scanners and a tape, some of the numbers the machine can, but not others, and not all of them." (McCulloch & Pitts, 1943, p. 129).

The theoretical contribution of the article is therefore that, when identifying the central nervous system with a logical machine, its behaviour must be characterized through the computation of classes of functions that are to be situated between the trivial combinatorial functions, computable by simple Boolean logics, and the too general class of functions, the recursive functions, which are computable by a Turing machine. Both authors defend a notion of computability that is proper to finite automata (Dupuy, 1985, p. 106). Although in the article in question they were in the first place concerned with the relation between mind, central nervous system and artefact, they actually laid down the foundations for the theory of finite automata.

As McCulloch and Pitts maintain that they bring a *psychological justification* for Turing's definition of computability and its equivalents, such as Church's λ or the recursive function of Kleene, and taking into account the specificity of a McCulloch-Pitts net with regard to finiteness, we can now understand why the identification between a logical machine and the brain was so very important for them. As the brain is material and finite, it is obvious that it cannot compute everything the machines of Turing can compute.

We have mentioned above that McCulloch, in his interpretation of the 'embodiment', reasoned in an idealized and theoretical way. This now turns out to be too general a characterization. McCulloch was concerned with the construction of artefacts that would allow for the incarnation of any theory of the 'mind' (this is what the theoretical question with regard to 'idealized' neurons amounts to), provided that certain factual aspects, aspects of causality and of finiteness, in this artefact should be taken into account (this is where the importance of the experimental study of mental and cognitive phenomena becomes visible).

The specific position of McCulloch in the first cybernetics can be shown through the comparison with another basical text, namely 'Behaviour, Purpose, and Teleology' of Rosenblueth, Wiener and Bigelow (1943). 'A Logical Calculus of the Ideas Immanent in Nervous Activity' can be seen as a radicalization of this text and that in two different ways (Dupuy, 1985, pp. 24 ff).

The first radicalization is that, whereas Rosenblueth et al. refuse to assign any reality to the 'mind' and regard every reference to this sort of properties to be a facile solution, McCulloch on the contrary wants to discover the material, but above all the logical mechanisms that 'incarnate' this mind. What he is really concerned with is the identification of properties of mind with the logical capacities of an artefact. But also we have to do here with an attempt to identify psychic phenomena with neural phenomena. McCulloch had in mind a study of the behaviour of natural phenomena inside the brain, via the logical operations of the artefact. This is not a return to some sort of mentalism, but rather a radicalization of the internalist position that was also to be found in the first article.[10] McCulloch only situates the study of the behaviour of natural phenomena within the brain: the formal neurons are seen as operators which transform inputs into outputs.

The second radicalization concerns the relation between *organisms and machines*. McCulloch radicalizes this relation by means of a double identificatory relation.

Wiener saw cybernetics as the science of the *analogies* between machines and organisms. McCulloch doesn't think in terms of analogy any longer, but in terms of identity: organisms *are* machines. As Dupuy remarks, this is not only a difference in point of view. Wiener is thinking as a mathematician who is looking for applications. McCulloch is at this point reasoning ontologically. Moreover, he has another conception of what the machine really is. For him, a machine is something of a logico-mathematical nature incarnated in the organism. He thinks of it as a 'logical machine', but at the same time one could say that, from his point of view, we have to do with a 'natural machine': nature and logics are identical. This is in clear contrast with Wiener, for whom the machine always remains an artefact and never loses its artificial character.

That McCulloch wanted to bring the psychological justification of Turing's definition of computability, and that he believed that the organism can be identified with the artefact, implies that he is important both on the epistemological and the physiological level. To posit that the brain is a computation-machine, comes down to explaining about the relations between mind and matter by accepting an intermediary reality, a reality which both relates to mind and to extension. The idea of a mind as a computing machine was not new at all; the identification of the nervous system with a computing machine was, on the contrary, very new and was for the first time concretized with McCulloch (Lévy, 1985a, pp. 214-215). McCulloch provides an extension of the thesis of Turing by posing that the brain is a computing machine; he goes beyond the idea of an equivalence between thinking and computational operations.

The notion of the brain as a digital computer, therefore, is present in a much deeper way in McCulloch's work than in that of Wiener. For McCulloch, cybernetics is the apotheosis of physics. It has to answer the question whether theoretical physics and knowledge in general are discussable in neurophysiological terms. Cybernetics has to help physics to deliver its masterpiece, and prevent it from falling into metaphysics.

Hence, we can say that the position of McCulloch with regard to cybernetics in general was a pretty radical one in as far as relations between machine, mind and organism were concerned. But there was another way in which he was clearly ahead of first cybernetics. In trying to understand the question of autonomy via a network, he arrived at insights which already belonged to second order cybernetics. Moreover, the alternative model-conception, which gradually took shape starting from these insights, is quite essential to argue the special statute of his experimental epistemology.

3. Heterarchy and autonomy: the anticipation of an alternative model-conception

McCulloch attempted to give the notion of a hierarchy of goals and values, the notion of a choice out of different goals, a place in a net of formal neurons. He realized that aspects of utility and control in the cybernetics of Wiener and Rosenblueth were essential to fix and select goals. Nevertheless, what he was after was more than the mere reduction of the goal conception. His interest in the relation between organism and machine certainly plays a role here. He wanted to know more about the way in which this aspect of "utility is mediated specifically in the physiology of particular anatomical structures" (McCulloch, 1965b, p. 365). It is clear that in this context the selection of a goal is not so evident. In this regard he arrives at the notion of a heterarchy of values.[11]

We will give an example of this with his 'reverberating circuits'. We shall restrict ourselves to his interpretation of abstract ideas. (McCulloch, 1943, 1945, 1965c, 1965d, 1965e).

Ideas or concepts are described by McCulloch as abstractions, as invariants, belonging to a level that is superior to the one of perception, but also marked by a fundamental independence with regard to the form that is perceived. For a network, in order to have the disposition of abstract ideas, the presence of closed circuits suffices, enabling the encoded invariance to circulate infinitely. There are two properties of networks which enable the reference to the past to be indefinite: the loops or reverberating circuits, and the fact that several initial states can lead to the same output.

Thus, the net consists of different elements of which some are organized in a loop. A series of signals realizing a certain configuration, can be reverberated by these loops, which is not the case with nets without circularity. "In a net in which there are no re-entrant paths a signal anywhere in the net implies a signal in a neuron nearer to receptors, and so backward in time until we arrive at the receptors. The signal here and now in this net implies the signal sent there just then. But once set going, a disturbance in a closed chain implies that there was a signal in its input at some time but does not indicate at what time. In short, the reverberating activity patterned after something that happened retains the form of the happening but loses track of when it happened." (McCulloch, 1965e, p. 77)

At the level of consciousness we encounter the implications of this in the form of a gap between intention and action. For McCulloch, consciousness boils down to the 'idea of the idea', organized in a circularly defined network. Although we are not cut off from perception because of the mechanism of abstraction, there is still a specific degree of freedom, a neurologically recognized gap between perception and idea, between intention and action. "Consider for a moment a computing machine in which there are no closed paths, that is, no circuits around which signals may chase their tails. (...) Their signals likewise imply the world impingent upon its sense organs. In the strictest sense of the word, what goes on in such a nervous system implies the world impingent upon these sense organs. *But note that the domain of implication extends only backward in time.* Even if the threshold of every relay were fixed, between the moment at which a given set of impulses started from a spot in our brains to our hands and feet, there might intervene other impulses coming, by shorter paths from the outside world, to our effectors. Thus, in the forward direction, the relation falls short of implication inasmuch as aught else intervenes. In short, *our thought does not imply our action but, as we say, only intends it. Perforce we distinguish between futurity and intention. Our notion of our will has, I believe, arisen from this enforced distinction, and its perennially questioned 'freedom' presumably means no more than that we can distinguish between what we intend and some intervention of our action. If I shall do what I will do, then my will is free.*" (McCulloch, 1965d, p. 149, our cursivation) Prediction is possible, abstraction is possible, but not without the loss of the specific historical way that has led to the present actual state, and not without sacrificing the impact of ideas on reality, on action. Because the history of an idea can never be recovered, the idea itself cannot have a decisive impact on a specific action. This is what causes the gap between intention and action and that's how the 'illusion' of freedom is created.[12]

Thus McCulloch illustrates that the indefiniteness of the reference to the past forms an anti-pole, a counterweight for abstraction. *The indefiniteness of the reference to the past is the counterweight which inevitably goes together with the abstraction by which our knowledge is made useful.* The consequence of this is that memory, although it is supposed to have started from a specific point, cannot return to the originating point of a specific configuration. Although a circular network is determined by its present state and by its input in a strictly deterministic way, it is absolutely impossible to reconstitute the past state from the present one.

Even though this fundamental idea has been repeatedly formulated in the history of philosophy, McCulloch gives it a completely new interpretation, leading to a possible concretization in the artefact. We have emphasized in relation with his net of formal neurons that he left room for the dimension of the possible, for the experimenting within the confines of the ideal. Although McCulloch starts from an identificatory relation between artefact and organism, the difference between natural and artificial can and should be made problematic on the basis of his insights. Through the artefact we see that the statute of the modelling is shifting. What we are dealing with here is the distinction between classical model and model via the artefact. In the 'experimental' epistemology of McCulloch we are mainly confronted with the model in the latter sense. We could posit that the artefact, when defined in a formal and idealized way, gains in autonomy after the first cybernetics. In a certain respect one could even say that it also loses part of its 'artificial' character, or at least, part of the properties which are traditionally connected with the artefact. The borderline between natural and artificial becomes vague, but at the same time the knowability of the artefact has decreased. When we inquire into the modelling statute of an 'autonomous' artefact, we observe that the only thing which remains is that the network, at the behavioural level, is comparable with ways of behaviour of certain natural systems. There is no question any longer of reduction of purposive or cognitive behaviour (in the sense of the first text of cybernetics, 'Behaviour, Purpose and Teleology'). We could say, if we want to put the case quite strongly, that the problem of cognition is no longer put in terms of reduction, but in terms of imitation. The dimension of explanation (supposing such a dimension is connected with it) has to be judged in these terms.

4. Between possibility and finiteness: a minimalistic interpretation

From what is said above it is clear that it would hardly make any sense to inquire into the statute of the McCulloch-Pitts hypotheses, neither in the light of the tradition prevailing at that time, nor in the light of the current neurophysiological tradition.[13] We can't free ourselves of the impression that the question about the degree of obsoleteness or of having-been-falsified, is not a fundamental question with regard to their work. McCulloch's way of approaching cognitive phenomena remains important — despite the fact, for instance, that an actual neuron is not a McCulloch-Pitts neuron — and is more and more articulated, confirmed, inquired into, taken seriously. This sort of discussion as to the pertinency in the light of neurophysiological data would not only lead us too far, but there would also be some risk that the central question, namely the one with regard to the relation between experimental and pure epistemology, would be relegated to the background. What we aim at is to inquire precisely into the fact how the question of finiteness and the idealization of the network can be reconciled with each other in an abstract way, and how the specific importance of experimental epistemology can be linked to this.

Putting too much emphasis on the article of 1943 may lead all too easily to a static description of the work and significance of McCulloch, to a characterization in terms of 'idealized, top-down, atomistic', etc. The way in which his work has been received also bears witness to a one-sided and partial emphasizing of a formal idealized net of neurons which actually only models the 'mind' and not the 'brain', skipping chemical and biological steps and in so doing does not attempt to render its whole rich complexity. For instance, the difference to which Rosenblatt draws the attention, the difference between his own approach and the one of McCulloch and Pitts, is that the latter depart from the top and analyse in downward direction (top-down), from the ways of behaviour which a net *must be able to realize*, deducing the structure from it. Rosenblatt reproaches them for indulging in biologism and calls them monotypical. He proposes a genotypical approach on the basis of a class of networks of which the structure is incompletely defined (bottom-up).14

As a matter of course we should not underestimate the fact that McCulloch's model, in spite of the identificatory relations which he subscribes to, is an *idealized net of neurons*. McCulloch repeatedly emphasized that the mechanisms which he proposed were *sufficient*, but not at all *necessary* to account for the properties of mind. The mechanisms were *possible*, according to him, in view of the available anatomical and physiological knowledge, and because they enabled the network to produce a number of the properties required. He writes in 1947: "To demonstrate the essential consequences of known characters of neurons, any theoretically conceivable net embodying the possibility will serve." (McCulloch, 1965c, p. 46).

On the other hand, we have underlined above that McCulloch considered the aspect of finiteness to be of importance, and that in this respect he wanted to bring the psychological justification of the Turingian definition of computability. Finiteness implies the setting up of experiments, the concretization of hypotheses with regard to properties of the formal network. McCulloch was, in his approach of the scientific theory, clearly Popperian.[15]

We want to confront the two aspects by calling the position of McCulloch with regard to the question of finiteness, and more generally with regard to the search for the empirical roots of knowledge as it found its place in experimental epistemology, *minimalistic*.[16]

It is quite clear that the net of formal neurons makes the idea of the brain as a logical machine more accessible, more concrete, and to certain extent also more plausible. It is also clear that the theoretical interest of McCulloch is limited by certain empirical-concrete aspects, certain requirements which the artefact must fulfil in a minimal way, if we want the identification between machine and organism to have a certain plausibility.

The interest for finiteness and for certain empirical factors does *not* imply though that a complete neurological or neurophysiological adequacy is considered to be primordial. That's precisely the reason why we want to characterize the position of McCulloch and the statute of experimental epistemology as situated between a classical or purely epistemological approach (inquiring into the logical origins of knowledge, and not into the psychological, neurological or biological origins) and a practical, psychological or neurological approach (in which the idealized aspect with regard to questions of knowledge is lacking). Without being a modelling of the brain, his approach permits us to formulate specific hypotheses with regard to the nature and functioning of the brain.

Minimalism, as we see it here, is quite apparent from McCulloch's following statement: "We have focussed our attention on particular hypothetical mechanisms in order to reach explicit notions about them which guide both histological studies and experiment. If mistaken, they still present the possible kinds of hypothetical mechanisms and the general character of circuits which recognize universals, and give practical methods for their design." (McCulloch, 1965c, p. 65)

Experimental epistemology can, within such a minimalistic interpretation, be a guideline, a generator of hypotheses and ideas, and also a constraining factor in scientific investigation: experimental epistemology departs from some concrete and empirically based limitations, and within these general epistemological ideas can expound. In principle, such a form of epistemological minimalism does not exclude the idea of a reduction.[17]

What is the advantage of a minimalistic approach ? Minimalism is a way of providing experimental epistemology with a practical, scientific and, in spite of that, a fundamentally epistemological statute. Whereas traditionally one quickly encounters difficulties in defining the exact epistemological statute of experimental epistemology (with reference to the fact that epistemology is both to be situated at a meta-level and should tend to have as much as possible a scientific character; also with reference to the fact that, though the borderline between science and epistemology becomes ever smaller, a fundamental distinction between epistemology and science must be kept up, as epistemology is a reflection concerning the objects which science investigates[18)], the minimalistic view on experimental epistemology has the advantage that it can escape this dilemma. Experimental epistemology does not seek to answer the question of the theoretical conditions of possibility of scientific knowledge — what Piaget called the proper constitutive constitution (but of course he considered epistemology to be in the first place a theory of knowledge).[19] On the contrary, the experimental epistemology indicates a.o. the concrete *conditions of impossibility*, the beacons beyond which there cannot be any question of knowledge at all, and enables the construction of pure epistemological theories which are compatible with these data.

The comparison between genetic and experimental epistemology can make our point still more explicit.

Piaget situated genetic epistemology in between general epistemology and genetic psychology. Genetic epistemology was directed at enriching general epistemology by reflecting on the notion of development. The investigations of his research centre constantly appeal on the experiment; an experimental verification is necessary. It is this aspect which differs from traditional epistemological investigations.[20]

McCulloch too reacted against an epistemology that would only be analytical and critical — his plea for an idealization and for a theoretical approach of cognitive phenomena must certainly be seen as a dismissal of the former interpretation of epistemology (Papert, 1965, pp. xiv-xv). We nevertheless have to see the statute of his experimental epistemology as fundamentally different from the one of genetic epistemology. The experimental basis is 'constituted' with McCulloch via the artefact, and this basis, within certain limits, renders an epistemological idealized reflection possible. The behavioural evidence, taken in a radical, i.e. 'McCullochian' sense, is still seen by McCulloch as a guarantee for a certain similarity between the artefact and what can correspond to it within the natural. Although the behaviour of the network is actually a hypothetical behaviour, derived from suppositions in connection with the system, rather than a description of certain observed behaviour (Lashley, 1965, p. 100), he nevertheless doesn't question, at that moment, the modelling statute of the artefact. However, specific problems arise around *modelling*, problems which for instance didn't present themselves in the same way in the genetic epistemology of Piaget. The special importance of McCulloch lies in the fact that via his experimental epistemology he anticipates a radically alternative conception of what a model is, a conception which is being developed by the study of autonomous systems on the one hand, and by an idealized representation of (autonomous) artefacts on the other.

We could put the statute of Piaget's genetic epistemology in the following series: experimental psychology, genetic epistemology (experiment in the traditional sense of the word), pure epistemology.

With McCulloch the series would be the following: neurology and neurophysiology, experimental epistemology (role of the artefact and the ensuing complexification of the model), pure epistemology.

The special role of the artefact with McCulloch has clearly a *liberating effect*, an effect also mentioned by Papert. Such an approach, as he puts it, allows for more hypotheses because it leaves the influence of detailed descriptions, neurologically as well as biologically, out of consideration. The experimentalism in the genetic epistemology of Piaget, though for several reasons it can't compare with experimentalism as it is to be found in the exact sciences, it has often been used as an argument for its 'objectivity'. The experimentalism of McCulloch has a quite different meaning. For one, it is much more disconnected from neurological data, and this by definition, as the artefact plays a completely different role in his approach. In addition to that, McCulloch illustrates via this artefact that there must be some sort of liberation from any experimental basis: the underdetermination of Quine is clearly to be traced back in this work.

What Papert does not notice is that we can find with McCulloch one of the basical points of minimalism, namely the going out from *some beacons, some ultimate limits* inside of which theory-building in general (not only the epistemological one) is possible. We believe that in this respect Papert characterizes one of the central features of McCulloch's approach wrongly as a "rational quest for necessity" (Papert, 1965, pp. xvii-xviii). McCulloch is looking, as he puts it, for the sufficient, but not for the necessary conditions. The minimalistic interpretation we defend here aims at showing that his experimental epistemology, starting from a number of conditions of impossibility (the ultimate limits), is a heuristical basis for the construction of an idealized, but probable theoretical epistemology.

By way of conclusion we want to indicate briefly some supplementary advantages of such a minimalistic position.

First, via the minimalistic interpretation of experimental epistemology one can avoid the skepticism which is to be found today on the radical constructivist side.

Secondly, one can also invalidate the antinomy which is argued by Cherniak in relation with the naturalistic position (see Van de Vijver, 1989). In *Minimal Rationality* Cherniak comes to certain conclusions with regard to the programme of naturalized epistemology, which we could have taken as a starting-point for the positioning of the experimental epistemology of McCulloch. For the tradition of cybernetics is not only to be inscribed in the naturalistic approach of knowledge as Quine accepted, but it has also given a new impulse to it (Guéry, 1989). As a matter of fact, this could already be observed in the quotations that appear at the very beginning of this text.

What Cherniak wants to argue is that the programme of naturalized epistemology implies an antinomy: even when skepticism is put aside, naturalized epistemology still provides the basis for a new conclusion, boiling down to the fact that there are serious limits to empirical knowledge. Nevertheless, although finiteness implies that there is a practical limit to our cognitive abilities, this does not mean, according to Cherniak, that the fundamental possibility of theoretical epistemological questions is ruled out. It is quite possible that practical limitations are not retained in all cases, for instance, with a Turing machine. Cherniak also underlines the fact that one of the most controversial parts of naturalism has to do with the fundamental epistemological situation of actual organisms with limited cognitive resources.

Through the minimalistic interpretation, problems in connection with verification are avoided, problems which Cherniak uses to show that a naturalized epistemology ultimately cannot dispense with a theoretical, purely epistemological interrogation. Concluding from our interpretation of experimental epistemology, it is directed at revealing limits, 'critical masses' as Cherniak calls them, in different systems, natural as well as artificial.[21] Experimental epistemology thus proves to have liberated itself from the exclusive concern of human systems with limited cognitive resources. This we call, in a certain sense, a point that has been scored.

* Senior Research Assistant of the Belgian National Fund for Scientific Research.

Notes

1. Following Heidegger, one can speak about a de-humanization by a synthetic a priori without subject, or about a 'deconstruction of the metaphysics of the subject' (Dupuy, 1985, p. 35). Ricoeur spoke, when describing the work of Lévi-Strauss, about a Kantianism without a transcendental subject'. This should not surprise us, in view of the relationships that exist between structuralism and cybernetics.

2. The most important of our informatory and critical sources with regard to McCulloch are: Lévy, 1985, 1985a; Palm, 1986; Papert, 1965.

3. A good survey of the evolution of his work is to be found in 'What Is a Number, that a Man May Know It, and a Man, that He May Know a Number ?' (McCulloch, 1965a). McCulloch started from mathematics, then studied psychology, and finally took up the study of medicine and of neurology in particular. His interest for theoretical biology made him partly an exception among the first cyberneticians. It is in the first place John Von Neumann who had McCulloch's work taken up in the world of biology.

4. Which brings with it certain advantages: "Au naturel le monde suffit à peine, à l'artificiel convient un espace clos" (Robinet, 1973, p. 32).

5. See McCulloch, 1965f, p. 163.

6. These psychons could either occur or not (on- or off-state), they were determined by former states of other psychons, determined in turn other psychons, and could finally be compounded into complex propositions (McCulloch, 1965a, p. 8).

7. The rules of connection have to do with the way in which neurons are connected: many input/single output, the negative or positive character of the input, the direction of their course, ... The rules of operation refer to the computation of the internal state of the neuron on the basis of the afferent input lines, of their exciting or inhibiting character and their threshold values. The value of the output, either 1 or 0, is determined by comparing the result of the computation with the threshold value of the neuron.

8. For an extensive discussion of the proof, see Von Foerster, 1984.

9. We won't go into the difference between 'embodied' logic and 'embodied' computation. We refer for this to the article of Daniel Andler in the present collection, who shows quite clearly in what sense this difference is of importance for problems of representation.

10. For the internalism we refer to Woodfield, 1976, pp. 160-201. For a more detailed discussion of the internalism in cybernetics: Van de Vijver, 1991, *Van cybernetica naar connectionisme. Een epistemologische studie van doelgerichtheid*, Ghent, Academia Press.

11. The concept 'heterarchic' is put by McCulloch in opposition to hierarchic. It refers to the circularity that is built-in in the network, making the existence of a highest and potentially controlling level impossible. This interpretation is clearly at variance with the prevalent interpretation of the term.

12. That this may form a point of discussion is quite clear. That the abstraction for a part takes its toll, should not necessarily entail that the history dimension loses *every* function. Nor does the fact that this function cannot immediately be recognized necessarily imply its non-existence. The fact whether or not one can recognize it can depend for instance on one or other specific mechanism which can show what this history has been, or which function it still has in the present. To formulate it in other words: the history can become 'sub-symbolic', or else is relegated to the 'unconscious', from where it can continue to have its effect, possibly unrecognized, until it becomes clear in the repetition that something else than the present constellation of factors is playing a part, until it becomes clear from the constantly

occurring difficulties that the past is not merely 'gone'. One can see that this presupposes a.o. a more complex view of the psychism, a view which in actual fact takes account of the different dimensions of time in a more complex way. If one should extrapolate McCulloch's theoretical position to practice, to the praxis of the psychiatrist for instance, then a history-dimension would never be required, neither for diagnostics nor for treatment. The aversion of McCulloch for psychoanalysis is quite noticeable. The same idea is also to be found with Ashby and with several representatives of the systemic therapies.

13. The formal neurons of McCulloch clearly do not correspond completely to real neurons, yet the idea is taken over that a neuron is an element of computation with algorithmic procedures operating on it, though of a much greater complexity than a McCulloch-Pitts neuron. Current neurobiology therefore is certainly indebted to McCulloch, as in neurobiology the brain is seen as an information-processing system (Cf. Lévy, 1985, p. 251). The discussion following on 'Why the Mind is in the Head', during the Hixon Symposium, clearly maps out the distribution of the positions with regard to this (McCulloch, 1965e, pp. 87-141).

14. Rosenblatt, 1962, p. 386. P. Livet also speaks about a prospective versus a retroactive approach (Livet, 1985, p. 148).

15. Even though he never is seen to quote this author (Lévy, 1985, pp. 243-244). A text that clearly bears witness to the importance of the testability is 'How we Know Universals', both in its general design and by the fact that the data with regard to vision, to McCulloch's great pleasure, were later falsified.

16. The use of the concept 'minimalism' was inspired by C. Cherniak's *Minimal Rationality*, 1986.

17. Seeing that minimalism can always be considered as a provisional solution which we have to accept within the scientific context that as yet remains incomplete. McCulloch seemed to be of this opinion: "We have stated our problems exactly and either solved them or seen how they can be solved. But you may have noticed that we have often made nervous nets to fit to the facts instead of seeking them in the nervous system. We have shown that a net can be made willfully to embody this or that idea, and to learn to embody this or that idea, but *not that these nets are even likely to exist.* Artificiality cannot be held against a man-made computing machine, but it is too much to have to suppose that each of ten billion neurons is connected according to any complete set of drawings and specifications." (McCulloch, 1965g, p. 270, our cursivation).

18. According to Blanché, epistemology remains a secondary démarche with regard to science, a reflection on science. Here, we have to do with a difference in linguistic levels. The epistemology must always take care "à faire totalement abstraction des

choses que vise la science qu'elle prend elle-même pour objet, en d'autres termes qu'elle s'assigne comme domaine exclusif d'étude, non pas ce sur quoi porte la science à laquelle elle a affaire, mais seulement ce qu'elle en dit. Sous peine de confusion entre les niveaux du langage, elle doit donc considérer uniquement le *discours* scientifique, traité comme un système de signes se combinant entre eux selon certaines règles, indépendamment de ce qu'ils peuvent évoquer." (p. 120, cursivation by the author). Or else: "N'exigeons pas de l'épistémologie un degré de positivité qui dépasserait celui de la science sur laquelle elle porte." (p. 124). On the other hand, epistemology cannot be compared to philosophy. This becomes quite clear when Blanché explains that a.o. Bergson and Ruyer did not provide an epistemological contribution to biology, but rather a philosophy of life.

19. It is quite easy to understand that Piaget considers epistemology and theory of knowledge to be synonyms. "C'est que la science et l'esprit scientifique, aussi bien dans l'évolution des sociétés que dans le développement de l'individu, se consituent progressivement, sans jamais parvenir à un état d'achèvement. Dans ces conditions, toute épistémologie génétique, qu'il s'agisse d'histoire des sciences ou de psychologie de l'enfant, s'élargit nécessairement en une théorie de la connaissance, puisqu'elle se propose de parcourir tous les stades par lesquels on parvient à ce que nous tenons aujourd'hui pour la connaissance scientifique — c'est-à-dire de considérer la connaissance sous des formes que nous ne pouvons juger que comme préscientifiques, et auxquelles cependant nous ne pouvons dénier toute valeur cognitive puisqu'elles préparent les progrès ultérieurs." Blanché, 1983, p. 15.

20. We won't go into the fact that this experimental basis has always been, in a strange way, the weakest point in the programme of Piaget. This doesn't mean though that it is not an essential question, as we believe that the introduction of the dimension of development certainly has relevance to this. Blanché, 1983, pp. 39 ff.

21. "... the recent work in the field of computational complexity raises the possibility that there may be another 'critical mass' for a knowledge representation, a maximum size threshold above which belief systems must in effect disintegrate." C. Cherniak, 1986, p. 128. This is an example of a typical achievement of experimental epistemology.

References

Blanché, R., 1983 (1972), *L'Epistemologie*, PUF, Paris, Coll. 'Que sais-je ?'.

Cherniak, C., 1986, *Minimal Rationality*, MIT, Cambridge, Massachussets.

Dupuy, J.-P., 1985, L'essor de la première cybernétique, *Cahiers du CREA*, Paris, nov. 1985, no. 7, pp. 7-140.

Guéry, F., 1989, Une "Philosophie de la Nature" aujourd'hui, in *Encyclopédie Universelle*, A. Jacob (dir.), vol. 1, pp. 454-461.

Lashley, K., 1965, Discussion following 'Why the mind is in the Head', *Embodiments of Mind*, Cambridge, Massachussetts, pp. 100-102.

Lévy, P., 1985, Brèves indications sur les vies de Warren McCulloch et de Walter Pitts, *Cahiers du CREA*, no. 7: Histoires de cybernétique, nov. 1985, pp. 203-211.

Lévy, P., 1985a, L'Oeuvre de Warren McCulloch, *Cahiers du CREA*, no. 7: Histoires de cybernétique, nov. 1985, pp. 211-257.

Livet, P., 1985, Cybernétique, auto-organisation et néo-connectionisme, *Cahiers du CREA*, nov. 1985, no. 8: Généalogies de l'auto-organisation, pp. 105-155.

McCulloch, W.S., 1945, A heterarchy of values determined by the topology of nervous nets, *Bulletin of Mathematical Biophysics*, vol. 7, pp. 89-93.

McCulloch, W.S., 1965, *Embodiments of Mind*, The MIT Press, Cambridge, Massachusetts, pp. 402.

McCulloch, W.S., 1965a (1961), What Is a Number, that a Man May Know It, and a Man, that He May Know a Number ?, in *Embodiments of Mind*, The MIT Press, Cambridge, Massachusetts, pp. 1-19.

McCulloch, W.S., 1965b (1964), A Historical Introduction to the Postulational Foundations of Experimental Epistemology, in *Embodiments of Mind*, The MIT Press, Cambridge, Massachusetts, pp. 359-373.

McCulloch, W.S., 1965c (1947), How We Know Universals: The Perception of Auditory and Visual Forms, in *Embodiments of Mind*, The MIT Press, Cambridge, Massachusetts, pp. 46-67.

McCulloch, W.S., 1965d (1948), Through the Den of the Metaphysician, in *Embodiments of Mind*, The MIT Press, Cambridge, Massachusetts, pp. 142-157.

McCulloch, W.S., 1965e (1951), Why the Mind is in the Head, in *Embodiments of Mind*, The MIT Press, Cambridge, Massachusetts, pp. 72-142.

McCulloch, W.S., 1965f (1955), Mysterium Iniquitatis of Sinful Man Aspiring into the Place of God, in *Embodiments of Mind*, The MIT Press, Cambridge, Massachusetts, pp. 157-165.

McCulloch, W.S., 1965g (1952), Finality and Form, in *Embodiments of Mind*, The MIT Press, Cambridge, Massachusetts, pp. 256-276.

McCulloch, W.S.; Pitts, W., 1943, A logical calculus of the Ideas immanent in nervous activity, *Bulletin of mathematical Biophysics*, vol. 5, pp. 115-133.

Palm, G., 1986, Warren McCulloch and Walter Pitts: A Logical Calculus of the Ideas Immanent in Nervous Activity, in G. Palm en A. Aertsen (eds.), *Brain theory*, Springer Verlag, Berlin, pp. 229-230.

Papert, S., 1965, Introduction, in *Embodiments of Mind*, W.S. McCulloch, The MIT Press, Cambridge, Massachusetts, pp. xiii-xx.

Robinet, A., 1973, *Le défi dybernétique. L'automate et la pensée*, Gallimard, Paris, pp. 232.

Rosenblatt, F., 1962, Strategic approaches to the study of brain models, in *International Tracts in Computer Science and Technology and their Application*, vol. 9: Principles of self-organization, H. Von Foerster and G.W. Zopf (eds.), Pergamon Press, Oxford, pp. 385-402.

Rosenblueth, A.; Wiener, N.; Bigelow, J., 1943, Behaviour, Purpose and Teleology, *Philosophy of Science*, jan. 1943, vol. 10, no. 1, pp. 18-24, reprinted in *Modern Systems Research for the Behavioural Scientist*, W. Buckley en A. Rapoport (eds.), 1968, Aldine Publishing Company, Chicago, pp. 221-225.

Van de Vijver, G., 1989, Self-organization and naturalism: perspectives and deadlocks of the new connectionism of Atlan and Varela, *Communication & Cognition — Artificial Intelligence*, vol. 6, no. 2-3, pp. 177-198.

Van de Vijver, G., 1991, *Van cybernetica naar connectionisme. Een epistemologische studie van doelgerichtheid*, Academia Press, Ghent.

Von Foerster, H., 1984 (1967), Computation in Neural Nets, reprinted in H. von Foerster, *Observing Systems*, Intersystems Publications, The Systems Inquiry Series, California, pp. 6-71.

Woodfield, A., 1976, *Teleology*, Cambridge University Press, Cambridge.

FROM PALEO- TO NEO-CONNECTIONISM*

Daniel Andler

U.F.R. de Philosophie
Université Charles de Gaulle, Lille III
&
Groupe de Recherche sur la Cognition, CREA
Ecole Polytechnique & CNRS, Paris

Abstract

The paper seeks to clarify some conceptual issues raised by contemporary connectionism, and its relation to classical computationalism, by means of a comparative study of the major features of early, 'paleo'-connectionism, and late, 'neo'-connectionism. An examination of the celebrated inaugural paper by McCulloch and Pitts (1943) reveals a deep ambivalence: their neural nets are presented both as 'logic machines' and as neural near-equivalents of Turing machines. It is shown that the lack of an intrinsic, 'embodied' representational level prevents these nets from attaining real logical powers, while as computational devices they herald contemporary computationalism in both classical and connectionist forms, but leave its representational face entirely in the dark.

Connectionism really gets off the ground with the first attempts at modelling perception, in the 1947 paper by Pitts and McCulloch on perceptual invariants, and in Rosenblatt's work on perceptrons. But it is only with the introduction of a representational level that connectionism breaks out of the narrow confines of perception proper: this, more perhaps than the well-known technical innovations, is the hallmark of neo-connectionism and may allow it to account for higher cognitive processes. However, the notion of representation at work in Parallel Distributed Processing(PDP)-style connectionism remains closely tied to the classical one, and thus unexplicated. The recent attempts, within the 'attractor neural net' school, to account for intrinsic representation, may give connectionism, of which the deep unity through changes is upheld, its own proper foundations.

Introduction

The very label 'connectionism' points to a simple question, whether or not

G. Van de Vijver (ed.), New Perspectives on Cybernetics, 125–146.

a prefixed 'neo' emphasizes it: what are the links and oppositions between the old and the new connectionism — between say McCulloch, Pitts and Hopfield, or Rosenblatt and Parallel Distributed Processing (PDP) ?[1] This paper will not be an exercise in "Contrast and Compare", nor will it attempt to list and relate all the main ideas and trends as they have historically emerged. Instead, it will focus on a small number of puzzles, such as the following, expressed in a deliberately provocative and somewhat unfair way:

- How can one seriously ask today whether nets can *really* do logic, while, as everyone knows, nets in McCulloch's days did exactly that (and only that) ?

- More generally, how is it that the problem(s) of cognition is (are) nowhere near solved half a century after they have been proclaimed solved 'in principle' by McCulloch (and Von Neumann), while neither of the main contenders to cognitive truth today — classical computationalism and connectionism — have unequivocally disowned McCulloch ?

- How could the mere adjunction of some 'hidden' units between the layers of the perceptron (actually envisaged by its presumed grave-diggers, Minsky and Papert), coupled with faster (Von Neumann) computers, change the face of connectionism, of AI, and perhaps even of cognitive science at large ?[2]

This paper therefore is not an attempt at serious intellectual history. It is an attempt at clarifying some issues in the foundations of contemporary cognitive science by examining certain aspects of earlier conceptual frameworks for the study of the 'mind/brain'. No doubt the latter enterprise would benefit greatly from progress in the former; the present paper will surely require revisions in the light of genuine historiography.[3]

1. Two perspectives on McCulloch and Pitts's neural nets

The harvest brought in by the great logicians of the 1930's was bountiful indeed. It included the notion of a formal system, of which the propositional calculus and the predicate calculus became the most important examples, and the notion of a recursive function, which provides the basis for a universal, completely general theory of computation. There are important links between the two notions, as was recognized from the beginning and has become ever more evident in the last twenty-five years or so. However, the basic fact remains that they are quite distinct. Not only are they *prima facie* as removed from one another as, say, the English language and the algorithm for long division, but it takes quite a bit of work to bring one to bear on the other. For example, to a large extent, Gödel's *tour de force* in his proof of the incompleteness theorem consisted in exploiting the computable nature of the morpho-syntactic operations in a formal system and the representability of recursive functions in any sufficiently strong formal arithmetic. This is one indication, among many, which should help the layman realize the distance separating the 'and' in such textbook titles as *Computability and Logic* (e.g. Boolos & Jeffrey, 1974) from any version, however weak, of the 'equal' sign.

In their famous 1943 paper, McCulloch and Pitts defend two claims. The first and presumably central one is indicated, if somewhat obscurely, by the title: "A logical calculus of the ideas immanent in nervous activity". The claim is that some neural nets 'do' logic, more precisely, propositional logic. The second claim, which occurs at the very end of the paper and comes, by way of proof, with a mere "It is easily shown that ...", is that a neural net can (more or less[4]) compute any number that a Turing machine can compute.

The resulting trace in collective memory is thus ambivalent: close one eye, and you see 'embodied' logic; close the other, 'embodied' computation appears. It is somewhat of a mystery to the present writer that, of the several commentators he has read (first and foremost the authors of the foreword to the new edition — Jerome Lettvin — and of the introduction to the first edition — Seymour Papert — of McCulloch's collected papers, and the editors of *Neurocomputing*), none has bothered to peel apart the two threads. Perhaps the distinction has been seen as too obvious to deserve attention — as indeed it is to anyone who examines the paper from the standpoint of logic.

However, one may suspect that some deeper confusion — or disagreement — is lurking, for example when one recalls one of Jerry Fodor's favourite mottoes: "No representation, no computation" (see *e.g.* Fodor, 1981, 1985; Fodor & Pylyshyn, 1988). Stripped from context, it seems to imply that representation goes hand in hand with computation. If you think of what is represented as, say, states of affairs, or propositions, then you are tempted to confuse calculus on propositions with general computation, and then perhaps just logic with computation. Then you are bound to equate 'embodied' logic with 'embodied' computation, and see *one* result in the McCulloch-Pitts paper where you should see two admittedly related, but conceptually distinct, results. And to confuse two very different notions of completeness or universality: one which applies to a set of connectives, such as {**not, and, or**}, and indicates the ability (of that set, or of any mechanism, say a neural net, which 'realizes' it) to express any complex proposition, the other which applies to a computing machine and indicates the ability (of that machine, or of any mechanism, say a neural net, which emulates it) to compute any function which can be computed, that is to provide the value $f(x)$ of any given computable function f on any given integer x belonging to the domain of f.

Tracing the root of this confusion will help us grasp what is perhaps the most important difference between the old and new connectionisms. So let us re-examine briefly the two theses put forward in the 1943 paper.

1.1. McCulloch-Pitts nets as logic machines

Take a neural net comprising at least three neurons A, B, C, and assume that A's being active means that **p** (a proposition), B's being active means that **q** (some other proposition), and that C is active except when A is active and B isn't. Clearly, then, that net, under the proper description, 'embodies' the logical relation '**p** → **q**'. Assume further that the brain comprises a structure which, for some purposes, is faithfully modelled by that neural net. Then you have an explanation of the brain's ability to effectuate logical implication; *ergo*, the *mind*'s ability to grasp the relation of logical implication between propositions is accounted for. Similarly, of course, for any other logical relation one might think of. "Thus the [...] formal aspect of that activity which we are wont to call *mental* [is] rigorously deducible from present neurophysiology." (McCulloch & Pitts, 1943, in McCulloch 1965/1988, p. 38)

The contemporary reader, duly warned by McCulloch himself ("Don't bite my finger, look where I'm pointing!"[5]), might as well pause at this point to catch his breath. For just about everything seems to be wrong with this line of reasoning. First, logic isn't just propositional calculus. Second, what the above net instantiates, at best, is not logical implication in general, but the complex formula **p** → **q**, with **p** and **q** fixed (they are the 'content' of the activity of neurons A and B respectively). Third, in order to be a plausible candidate for the job of 'embodying' **p** → **q**, node C should possess an intrinsic property making it specifically the embodiment of **p** → **q**, rather than that of any other complex formula built from **p** and **q**, such as the (truth-functionally) equivalent ¬**p** v **q** or perhaps even some non-equivalent formula such as **p** v **q**. For **p** → **q** itself, whether regarded as a platonic proposition or as a sequence of symbols, is intrinsically, *non-relationally* distinct from any morphologically distinct proposition or formula. But what property could do the job in our net is, to say the least, unobvious.[6]

The first two objections can be disposed of fairly quickly in this context. As for the first, it should be admitted that logic, ever since cognitive science came into being, has been reduced to propositional calculus (be it somewhat enriched, *e.g.* modal, or deviant, *e.g.* probabilistic, intuitionistic, etc.) — it would therefore be highly unfair to demand more of the pioneers than we generally demand of ourselves. Besides, McCulloch and Pitts did not lose awareness of the need for quantification; as pointed out by McCulloch (McCulloch, 1961, in McCulloch

1965/1988, p. 10), nets with circles can give rise to reverberation, which is a form or model of memory, which can in turn be seen as providing a temporal equivalent to existential quantification. This may be a bit swift, but allows us to at least suspend the objection.

The second concern could perhaps be alleviated as follows: after all, **p** and **q** cannot plausibly be thought of as completely fixed, else the net would be pretty useless: they presumably vary with time and circumstances (or at least their truth-values do, perhaps because they contain, be it implicitly, some indexical element: **p** might be "It is raining" or "It is raining here now"). But then why not let the contents of A and B vary completely freely ? Then the net would 'embody' $\mathbf{p} \rightarrow \mathbf{q}$ for *variable* **p** and **q** after all. Of course, this suggestion leads to the problem of distinguishing between the *activity* of neuron A and its *content*, which could be avoided as long as **p** was kept *perfectly* fixed — the activity of A *meant* **p** (or meant that the state of affairs expressed by **p** was 'believed' to hold by the net, or by the brain).

But this, in another guise, is our third objection. In contemporary cognitive science, we would say that A *represents* **p** and that therefore A being active means that **p** is deemed true by the system. We thus distinguish between the represented proposition and its assumed truth value, and ask first in virtue of what A represents **p** rather than **p'** (and also why A represents anything at all), and only second in virtue of what A is active when, and only when, **p** is in fact true. In the case of neuron C, we would want to be in a position to assert that it represents $\mathbf{p} \rightarrow \mathbf{q}$, and we might then want to explain why C is on precisely when, in virtue of the states of neurons A and B, $\mathbf{p} \rightarrow \mathbf{q}$ is true, by invoking the inner workings of the net ABC.

Now the question is whether there is anything in McCulloch and Pitts's avowed intentions, or in the structure of the nets themselves, that would allow us to make a similar move. In other words, is there some kind of representation intended or inscribed in the McCulloch-Pitts net ? With this question in mind, we find that there are two readings of the 1943 paper, which we now examine in turn. There is a completely harmless way of construing the 'immanent calculus': propositional logic *models* cerebral activity. Thus, in their introduction, the authors write: "The «all-or-none» law of nervous activity is sufficient to insure that the activity of any neuron may be represented as a proposition. Physiological relations existing among nervous activities correspond, of course, to relations among the propositions; and the utility of the representation depends upon the identity of these relations with those of the logic of propositions."[7] In other words, McCulloch and Pitts do for neural nets — and therefore, or so they hope, for the brain (under description) — what Claude Shannon had done for electric circuits in his master's thesis (Shannon, 1938) a few years earlier: use logic (actually, Boolean algebra) to describe or model the behaviour of a complex system. The success of the approach, in either case, is based on the empirical fact that the physical magnitudes which characterize the system are 'disciplined' by the *mathematical* laws or rules of propositional logic.[8]

Such a reading is harmless conceptually, quite precious, as it will turn out, in the development of the digital computer concept, and completely uninformative from the cognitive or informational standpoint: neurons, or neuronal activities, on this reading, do not represent. Instead, they *are represented* by propositions and their truth values — just as parts of switching circuits can be, and many other things in the physical world.

We will return to this perspective, but first another, more informative and less harmless reading should be envisaged. Sometimes McCulloch and Pitts seem to suggest that the «all-or-none» character of neuronal activity makes it 'inherently propositional', thus conferring upon "all psychic events [...] an intentional, or «semiotic» character" (McCulloch, 1965/1988, p. 37). It then sounds as if neurons were actually in the business of standing for states of affairs. Taken in isolation, out of context, such an idea would seem preposterous: the fact that sails come in white and non-white doesn't make them *inherently* propositional — it took a special convention between a father and a son to lead the former to believe, falsely as it turned out, that the blackness of a certain sail meant the demise of the latter. But Lettvin, in his foreword, provides essential background: "Once it was realized in the nineteenth century that nerve fibers conduct electric pulses and that the pulse trains on these fibers carry meaningful messages, the problem was to account for how such information was processed by the brain as nerve net." (McCulloch, 1965/1988, p. viii).

Thus neurons do after all represent, or process 'meaningful messages'. On that reading, there is, in the background, some theory of representation. But not only is it not provided; nowhere do McCulloch and Pitts show any sensitivity to the need to distinguish between the formal object **p**, or the 'intention', the '*Sinn*' on one hand, and the truth value or '*Bedeutung*' of **p**[9] on the other. There lies, as we will see, a major difference with neo-connectionism. But right now it leads us to ask in what sense then the first generation cyberneticians took neural nets to 'do' (propositional) logic.

Although the answer is quite obvious, let us take the historical route to (re)discover it. Lettvin provides yet another interpretation of what McCulloch and Pitts regarded the neuron as representing: he writes that "Warren and Walter were led [mostly by David Lloyd's work in 1939-41] to conceive of single neurons as [...] acting as gates".[10] But can a neuron represent both a proposition and a connective ? Perhaps neurons such as A and B above represent propositions, while C represents material implication (→) ? The answer is subtler. Let us state it first in approximate form, for the sake of clarity: a neuron is (or *embodies*) a connective, and its activity *represents* the result of applying the connective to the propositions which are fed to it — in other words, a neuron N embodies a certain function *f* and its being on or off represents the truth or falsity of the proposition *f*(**p**,**q**), where **p** and **q** are the propositions asserted by the activities of the neurons affering onto N.

Well, this is still not quite right, because *f* does not operate on *formulas*, but on 0's and 1's. So neurons embody not connectives, but Boolean functions, and operate not on formulas, but (at best) on truth values of formulas. Final correction: neurons being modelled (here) as threshold automata are not in themselves Boolean functions; these are realized by *nets* of formal neurons.

Thus we have finally reached the proper description of McCulloch-Pitts nets as logic machines: they embody, or realize, or instantiate Boolean functions. They should really be called Boolean machines. However, they can be regarded as performing 'blind' or 'animal' logic, provided an interpretation is somehow given to certain neurons, the 'peripheral' ones which receive 'information' from external sources; they can then be construed as *applied* Boolean machines.

What is missing, however, from this model of nervous activity in order to constitute a neurophysiological account of genuine logical thinking, or cogitation, is an internal system of representations — an inner blackboard ready to receive the material inscriptions of the formulas the truth values of which are being evaluated — and possibly others, kept in storage for future needs, for counterfactual reasoning, etc.

Incidentally, this model is not even a *plausible* neurophysiological account of blind logic, due to the random and unreliable nature of connectivity and single-neuron activity in the brain, as pointed out, in harsh terms, by second-generation connectionists such as Frank Rosenblatt (Rosenblatt, 1958, in A&R, p. 93, col. 2). Neo-connectionists will conclude that logical thinking does not occur at the aggregation or complexity level of single neurons, but (if it does occur at all) at a much higher level.

1.2. McCulloch-Pitts nets as approximations to Turing machines

All told, there appears to be little plausibility in construing the nets in the 1943 paper as logic machines implemented in an idealized brain. In fact, despite the title ("A Logical Calculus Immanent ..."), I used to think of that interpretation as bordering on slander, and only with reluctance came to admit that McCulloch at least did sometimes believe that he and Pitts had shown how brains, thus minds, do (propositional) logic.

If one looks at the technical setting of the paper — a search for a description of the behaviour of nets in terms of complex (quantified temporal) logical formulas, and conversely for nets of which the behaviour fits a given such formula — one gains a rather different outlook on their achievement. That this perspective is the one which considered scientific judgment imposes, is confirmed by comments made some time later by such thinkers as Von Neumann, Papert, and in fact McCulloch himself. Let us begin with the co-author of the famous paper:

"What Pitts and I had shown was that neurons that could be excited or inhibited, given a proper net, could extract any configuration of signals in its input. Because the form of the entire argument was strictly logical, and because Gödel had arithmetized logic, we had proved, in substance, the equivalence of all general Turing machines — man-made or begotten." (McCulloch 1961, in McCulloch, 1965/1988, pp. 9-10). Although I would rather not be called upon to fully explicate this statement, its conclusion is clear enough: the 'logical importance' of the 1943 paper is that the brain (or some subsystems of it) can be regarded as a Turing machine — in other words, as an 'embodied' computer, and possibly as a universal one, *i.e.* one capable of computing any (computable) function whatsoever.

Papert, in his preface to the collected papers (published under McCulloch's own supervision), at once confirms the primacy of this perspective, introduces an important proviso and points to the (otherwise evident) link between the two perspectives: "When we reach the end of the paper, we are rewarded for the effort [...] by seeing the first birth of a true mathematical idea: between the class of trivial combinational functions computable by simple Boolean logic and the too general class of functions computable by Turing machines, there are intermediate classes of computability determined by the most universal and natural mathematical feature of the net — its finiteness." (McCulloch, 1965/1988, p. xxvi). Let us disregard the proviso (as devoid of relevance for our present purpose) and simply stress that once it is recognized that the net as logic machine really is a Boolean machine or computer, it takes but one small conceptual (though technically steep) step to the idea of the net as a (quasi) Turing machine.

But with Von Neumann we now come to the heart of the problem posed by this sounder, mathematically richer outlook; in his address to the Hixon Symposium in 1948, he says: "[Their result] is that anything that can be exhaustively and unambiguously described, anything that can be completely and unambiguously put into words, is *ipso facto* realizable by a suitable finite neural network." (Von Neumann, quoted in Goldstine, 1972, p. 276). How are we to understand the '-thing' in 'anything' ? Obviously as a *process*. We are provided, according to Von Neumann, with a neurophysiological (or maybe just 'neural') account of all mental processes. However, Turing machines operate on integers (or perhaps on marks), while mental processes operate on practically any domain and actually rather seldom on integers (and never on marks, except in tic-tac-toe!). Well, that can easily be fixed: just code the objects of the domain — with the help of something akin to a Gödel numbering — and the perfectly specifiable process on these objects gets *ipso facto* transmuted into a (Turing) computable function.

Now what is such a coding ? It may come cheap to the mathematician, but the cognitive scientist, the psychologist-philosopher sees in it nothing less than a representation. And that — the notion of a representation — is left *completely* out of the picture provided by McCulloch and Pitts on this second reading: there is no trace of 'meaning' or 'information' left, no account of a mental *state*, nothing but the 'engine' driving the mind from state to state.

Of course, this is a typical case of retrospective historiography: the work of the past is seen as an incomplete puzzle, outlining the pieces to be provided by the next generation. Our purpose in this paper being to identify some fundamental differences between paleo-connectionism and later paradigms, this may perhaps be regarded as an acceptable way to proceed. For us, there is yet another interesting point. Everyone at the time had a clear notion of what was missing: McCulloch and Pitts had only provided an 'existence proof' (for any fully specifiable task, there exists a neural net that can accomplish it), and there remained to be built, for all the (important) cognitive tasks, actual nets conforming to the 'specifications'. The 'coding' seemed to pose no problem. Artificial intelligence will take the problem from there, inheriting a costly prejudice. It will in due course discover the paramount importance of coding, in other words, of 'knowledge'.

2. *Perceptual representation in paleo-connectionism*

Actually, it did not escape the first generation cyberneticians' attention that in one domain of mental activity at least, 'existence proofs' and other exhaustivity claims notwithstanding, all the work remained to be done. Not surprisingly, that domain was that of perception, which on one hand had, in the philosophical and psychological traditions, always been regarded as largely independent of cognition (of the higher cognitive processes), and on the other hand raised the problem of representation in a simple form.

What is quite a bit more surprising is to discover that in retrospect, it is only when it approached that problem that connectionism really got underway. The 1943 paper appears to today's connectionists as a false start, and the right track is taken with the much lesser known paper published in 1947 by Pitts and

McCulloch (in that order), under the title "How we know universals: The perception of auditory and visual forms", in the same obscure bulletin as their first paper (Pitts & McCulloch, 1947). "It is much more in the direction in which neuroscience and network research has progressed since the 1940's", write the editors of *Neurocomputing* in their introduction. And Rosenblatt, had he known about it (which is unclear as he does not list it among the references to his 1958 paper), would no doubt have approved of it, both because it deals with perception rather than logic, and because it takes as a major constraint the randomness, unreliability and partial indeterminacy of the nervous system seen at the neuronal level.

The link between the 1943 and 1947 papers is problematic. The second paper makes no mention of the first. But McCulloch, in the grand retrospective he sketches in 1961, clearly claims continuity between the two: "in our next joint paper ..." (McCulloch, 1961, in McCulloch, 1965/1988, p. 10), he writes, they showed how to deal with other invariants than time (memory, accounted for with the help of reverberating nets with circles in the 1943 paper, is to be thought of as a 'temporal invariant'). Upon examination however, there seems little, if anything, in common between the papers. Neural nets (in the 1943 sense) don't appear, neither do propositions and logic. The single common feature is the appeal to the idea that neurons, or neuronal nets, 'compute' values required by the hypothetical account of invariant perception; in other words, the authors operate in the framework of 'embodied' computation set up and illustrated in the 1943 paper. As a matter of fact, they attribute to real neurons and neuronal nets properties of formal neurons and neural nets.

These conceptual and/or rhetorical maneuvers should be much more carefully examined than I have the space here and competence now to do. Perhaps we should regard them for the time being as a tribute paid by logical vice to perceptual virtue, or by computation to representation. What seems beyond dispute, more seriously, is that the true function of the 1943 paper is to lay the foundations of computationalism, or 'embodied' computation, providing both a working hypothesis on the mind/body problem and the technical tools required for the conception of computing devices, "man-made or begotten". This conceptual and technical heritage is claimed by the classical and connectionist paradigms alike. The origins of the schism are to be found elsewhere: in the 1947 paper, for example, and generally wherever the emphasis was placed on perception and neurophysiology. In the actual flow of scientific information and influence, the most salient event relevant to the separation of the two main branches of computationalism came much later: it was the publication in 1958 of Frank Rosenblatt's seminal paper on perceptrons.

The difference in tone between the first McCulloch-Pitts paper and

Rosenblatt's is striking. Naturally, the fifteen years separating them, which saw the birth of the cognitive sciences in the contemporary sense and the beginning of their convergence, goes some way towards accounting for the difference. Cybernetics had shown the way, at considerable risk, and the prophet's stance, suited perhaps to McCulloch's situation and personality, would have been quite out of place in Rosenblatt's case. Similarly, clarity and simplicity of expression are easier to achieve in a somewhat settled intellectual landscape. However, this cannot be the whole story: the 1947 Pitts-McCulloch paper is closer in tone to Rosenblatt's than to the 1943 paper. At the same time, there is in Rosenblatt's paper an intellectual ambition comparable to the first, rather than the second of the earlier papers: the focus is on the abstract, completely general characterization of a basic cognitive task, and of devices capable of carrying it out — the 1943 and 1958 papers deliberately situate the main topic on what we would identify today as Marr's top, 'computational' level (Marr, 1982, pp. 19-31, 336-361). The 1947 paper is focused on the second, 'algorithmic', and third, 'implementational' levels.

The basic reason for the disparity, I contend, is that Rosenblatt's topic is easier than McCulloch and Pitts's in 1943 in one crucial respect: representation. Although Rosenblatt has no more of a general theory of representation than his predecessors, he is in a considerably better position than they are because he doesn't need one. To put it crudely, logic is symbolic, vision isn't.[11] Because duality of levels is essential to logic, and because the relation between levels involves a form of abstract representation, one cannot account for 'embodied' logic (or cognitive abilities with a logical dimension) without explicitly postulating an *embodied* representational realm — on pain of actually accounting only for embodied Boolean computation. On the other hand, one can set out to account for perception without any prerequisite notion of representation: insofar as the link between a given shape and its 'kind' (the class in which it is put by the system under study) can be given a purely (psycho)physical account, so that the 'kind' is a *natural sign* of the shape, the job of the theorist is done once he has established the psychophysical (or neurophysiological) chain between a collection of shapes in an environment and their respective 'kinds'. That job consists precisely in building a theory of a certain kind of concrete representation, and, to repeat, presupposes no general concept of representation.

I thus conclude, with no more than a semblance of paradox, that Rosenblatt's paradigm, just like McCulloch and Pitts's in 1943 (or 1947 for that matter), is computational alright but not representational. In Rosenblatt's case, there simply is no general notion of abstract representation, and this is as it should be. In McCulloch and Pitts's (1943), there should be one and there isn't. But beyond that difference, and the concomitant obvious one in topics, both theories bear that hallmark of paleo-connectionism: they leave embodied, material representation in the dark.

3. Computation and representation in today's competing paradigms

In this final section, my goal is to bring out some major differences between paleo-connectionism and neo-connectionism; it will however be useful to stop on the way in order to briefly examine the case of classical computationalism. The pace will be rather brisk, first because apt characterizations of the main contemporary approaches are readily available (see note 2), and second because of the retrospective historiography I have, for better or for worse, indulged in: more than once have I characterized a past state of affairs by pointing out what makes it different from our present outlook, or perhaps by describing it in contemporary terms, so that it appeared as a familiar figure with some important feature missing.

3.1. Classical computationalism

This mainstream approach (sometimes also called cognitivism) in cognitive science and artificial intelligence, from the mid-fifties to this day, rests on a specific way of articulating representationalism, a doctrine about cognitive and mental states, with computationalism, a doctrine about cognitive and mental processes.

Cognitive states in a (Von Neumann, AI-programmed) machine are data structures in specified locations; in a human mind, they are inner representations in the functional equivalent of specified locations. In both cases, the medium of the representations or data is an embodied formal language (a 'physical symbol system' in Newell and Simon's famous phrase (Newell & Simon, 1976)). Formulas of that language are endowed with a semantics, they are information-bearers, for example they express states of affairs or situations. They also have a form and lend themselves to syntactic manipulations.

This is where computationalism comes in. The syntactic manipulations turn out to be effective — they are among the 'exhaustively, unambiguously describable «things»' that Von Neumann talks about — in other words, they are, up to coding, computable functions. A Turing machine can therefore effectuate those manipulations, and an *embodied* Turing machine, or any equivalent physical computer, can perform on the *material* aggregates of symbols which constitute the cognitive states of a system the *physical* operations which correspond to the appropriate syntactic transformations. The semantico-syntactic parallelism characteristic of formal languages ensures that appropriate syntactic action leads to appropriate semantic content, so that for example the system is lead, in favourable cases, from true beliefs to true beliefs, or from desirable but distant things to still desirable but less distant things.

As already mentioned, McCulloch and Pitts provided, in their 1943 paper, the means to build computer components able to perform elementary computations, and therefore whole computers able to perform any complex (computable) operation — the functional similarity between these components and certain subsystems in the brain could (and can) then be exploited either to model the brain to account for its computational powers, or to build brain-like, or 'neuromimetic' computers — but the likeness to the brain is firstly a relative, secondly an optional matter. On the other hand, as the other reading of the paper shows, the intention was to apply the computational powers of begotten and man-made devices to propositions. For all these reasons, the 1943 paper heralds classical computationalism at least as much, if not actually more, than neo-connectionism. The absence of a representational medium makes it however a very incomplete prefiguration of classicism, and a close relative of both the Rosenblatt version of paleo-connectionism and also perhaps of certain minority, explicitly non-representational, new-cybernetic trends in cognitive science (see *e.g.* Varela, 1979/1989).

3.2. Input/output neo-connectionism

Connectionism nowadays is a convenient and somewhat misleading label covering a large variety of doctrines, styles and enterprises, ranging from theoretical biology to psychology, engineering-oriented AI, etc. Of the many important distinctions within this family of research programmes, only one need concern us here. A majority of connectionist models are essentially feed-forward neural nets with a distinguished layer of input units and another of output units. Most of the Parallel Distributed Processing, 'PDP' research[12] provides specific models of that kind, as well as a general framework within which to study them, but localist schools (see *e.g.* Feldman & Ballard, 1982) also produce input/output information processors. For the present purpose, I therefore propose the non-standard label 'input/output connectionism' as a convenient way of grouping together efforts aimed at producing or describing models of this kind.

These are in many respects the direct heirs to the perceptron.[13] On the other hand, they depart from the perceptron in several important ways.

The best known series of differences bear on the architecture of today's nets and on the almost endless variations affecting the range of activity levels and the transition law of the units, as well as the learning procedure. Nets nowadays have 'hidden' units which they use as they please, thus displaying a modicum of self-organization. They can assume a probabilistic character. They follow 'back-propagation' or 'simulated annealing' learning protocols, etc. These changes (together with enormously more powerful computing resources) have enabled the 'neo-perceptrons' to overcome some of the technical shortcomings responsible for the virtual disappearance of the original perceptron.

The major conceptual differences lie elsewhere. Input/output connectionism shares with, and in fact for the most part borrows from classical computationalism its notion of representation. Admittedly, the emphasis is often laid on the differences: connectionism wishes to reject the classical commitment to some 'language of thought' fashioned after the formal languages of logic, it also stresses the advantages and significance of distributed, versus local, representations. But this is not the place to dwell on these differences, and the fact remains that connectionism (of the input/output kind) is representationalist in basically the same sense as classicism.[14] With respect to paleo-connectionism, this is of no little consequence.

By conceiving their nets as embodying processes acting on (embodied) representations, contemporary connectionists extend the range of cognitive phenomena accessible to a connectionistic account or simulation far beyond perception. Just like the classicists, they can help themselves to *abstract*, and not only *natural* representations. The higher cognitive processes can now conceivably be accomplished by nets — while Rosenblatt lucidly admitted that "some system, more advanced in principle than the perceptron, seems to be required at this point." (Rosenblatt, 1958; A&R, p. 111, col. 1). Of course, serious doubts have been voiced about the actual prospects for extending connectionist accounts all the way to language and logical reasoning.[15] Of course, the mere adoption of a general notion of representation is not enough to yield a universal connectionist methodology even in principle applicable to all tasks: in fact, there is considerable disagreement on how the net can be fed the right representation (the critical suggestion is that the actual job of assigning the appropriate representations remains outside the scope of connectionism). But these considerations, however much they deserve attention in present-day research, do not detract from the importance of the conceptual jump which has allowed neo-connectionism to escape the prison of natural perception: guided by the classical view, it can now approach non-perceptual phenomena in a perception-like way.

As a second consequence, in conjunction with the more specific idea of distributed representations, PDP-style connectionism has developed a two-level, emergentist view of cognition, freeing the physical, computational level from the semantic, symbolic level.[16] This distinguishes PDP connectionism from classicism (at some risk, it must be admitted (see Petitot, 1990; Andler, 1990)), and also from paleo-connectionism. Of course, connectionism always distinguished the level of single neurons and local connections from the level of the whole assembly. But paleo-connectionism had no means to link one side of representation to one level, the other side to the other level. This move by PDP connectionism is most welcome. If successful, it could allow connectionism to finally enter the realm of higher cognitive processes through the front door. Logic at the neuronal level being incompatible with the neo-connectionist view of (formal and neurophysiological) processing, and remaining, as we have seen, but a gleam in McCulloch's and Pitts's common eye, only an emergentist perspective can hope to explain how a net can produce (real) logic. A similar approach is advocated by Smolensky and others for the case of language.

3.3. Attractor neural networks

Following the physicist John Hopfield, whose 1982 paper was the single most important factor in the renascence of connectionism, and to some extent the psychologist Donald Hebb, whose 1949 book was a major inspiration to paleo-connectionists already, a small group of investigators have begun developing different sorts of nets and using them in an attempt to rethink the most basic concepts of cognitive science or psychology.

No more than a few words can be said about this current, which, being rather less known and in some ways rather more difficult to understand than other brands of connectionism, would require far more space to introduce than I can take here.[17] I will content myself with two hints.

As machines, or physical systems, the nets under consideration (called 'attractor neural nets' or ANNs) are distinguished by fact that their connections are multidirectional (rather than feedforward, as in the input/output schools), and comprise no predefined input or output units. This makes them essentially autonomous dynamical systems, of which the behaviour can be described (and to some extent predicted) on the basis of their attractor landscapes: in the simplest case, the trajectory of such a system is akin to that of a billiard ball which is initially left to its own device, in a given position and with a given velocity, in a rugged landscape, and eventually finds a line of steepest descent which leads it into a point of locally minimal altitude (the simplest kind of attractor), where it rests.

Such a system needs to be interpreted in a novel way in order to be seen as a cognitive device. Hopfield's original idea was to interpret an attractor as a memory (in the psychological — as in 'fond memories' — not the computer science sense of the word): prompted by some stimulation which places it at a certain location in the landscape, the system eventually stops in an attractor, which is the memory evoked by the stimulus. In a similar vein, Daniel Amit has proposed to interpret the physical event of a rapid return to equilibrium as the correlate of significance: of all the perturbations to which the system is submitted, and launch it on a new trajectory, only those which rapidly lead to a new balance are significant, and their meaning or content is the attractor into which they have led the system.

Such a cursory description might suggest arbitrariness or fuzziness in a research programme which is singularly devoid of those traits. In fact, the mathematical apparatus brought to bear on the study of ANN's is quite powerful (more powerful by several orders of magnitude than anything McCulloch and Pitts, Rosenblatt or Minsky and Papert could have imagined) and new. Clearly this does not *guarantee* the soundness and depth of the programme itself, but it is an auspicious sign. Anyway, its seriousness is unquestionable, and the genuine question that needs to be asked at this point is how do ANNs extend or complete the itinerary which began with McCulloch and Pitts's 1943 paper.

This is my tentative answer. The first generation cyberneticians had overlooked the notion, and the problem, of representation altogether. Neo-connectionists first focused on the notion and helped themselves, with a hint from the classical computationalists, to a ready-made notion which remained (and remains) to be explicated (see *e.g.* Fodor, 1987). Now they seek an account of true, full-blooded, intrinsically intentional representation. Neural nets, concomitantly, started out as Boolean machines, then became non-classical computing devices operating on representations of non-classical medium, and are perhaps in the process of becoming autonomous dynamical systems endowed with intrinsic cognitive abilities dependent on an ecologically valid interpretation.

*

* *

The very last question may then be whether there remains anything at all of the initial inspiration. Quite a lot does remain, in my view. First, there is the spirit of the proposed solution to the mind/body problem, based on a mediation by the notion of a machine and that of functional equivalence of different machines — cybernetics' major contribution to the emergence of cognitive science (see Dupuy, 1985). Second, there is the consideration of 'begotten' cognition as the main source of knowledge and constraints: connectionist cognitive science has by and large remained a branch of neuroscience, and has never seriously considered relocating. Third, there is this humble but hardy device, the threshold automaton or formal neuron. Finally, there is the insight that robust collective properties of assemblies of simple components can result from incompletely specified, imperfectly stable, not fully regular or reproducible local characteristics; and that cognition, that robust property of central nervous systems, may thus emerge, phylo- and ontogenetically, from a hopelessly complex, ever changing bundle of neurons.

Could this be enough of a reason, as this paper comes to an end, to forget about 'paleo' and 'neo', letting connectionism stretch and progress, through false starts and false endings, from 1943 to this day ?

* The present paper grew out of talks given in 1987 in Jerusalem (Mishkenot Sha'ahanim workshop on "Minds and Machines", April), Cerisy-la-Salle (conference on "Approches de la cognition", June) and Ghent (conference on "Communication and Cognition - Applied Epistemology", December).

Notes

1. The word 'connectionist' is used by Donald Hebb in 1949 and by Frank Rosenblatt in 1958, as James Anderson and Edward Rosenfeld, the editors of *Neurocomputing*, point out. Since I delivered a talk under the title of the present paper at the Ghent conference, MIT Press published this most precious anthology (hereinafter A&R), and thus made the job of tracing the roots and paths of neural network research at once considerably easier, by providing, together with knowledgeable introductions, all the major papers, and quite a bit more difficult, by ruling out any simplistic history one might have been tempted to propose. One simple idea I have been able to retain is that of a three-stage development: the first period, fairly continuous, slow but extremely fertile, extends from the 1943 paper of McCulloch and Pitts's to Minsky's and Papert's book *Perceptrons*; it is followed by a transitory decade leading in turn to the present explosive stage (heralded by a number of bangs — the Hinton and Anderson collection in 1979, Grossberg's 1980 paper, Hopfield's 1982 paper, Feldman's and Ballard's 1982 survey). In this paper, 'paleo-connectionism' will refer to the first period, especially to the fifteen years 1943-1958, and 'neo-connectionism' roughly to the last twelve years, 1979-present. An important and difficult question is raised precisely by the transition and by the role played by investigators who didn't stop being 'connectionists' in 1969, nor waited until 1982

to become members. This question will not be addressed here, but it is likely that a full treatment of the questions which I attempt to address would benefit from and adequate account of the *interregnum*.

2. No attempt will be made to sketch its general framework, *viz.* cognitive science, classical computationalism and connectionism. The references are too numerous to list; for the classical perspective, see for example Haugeland 1978, Fodor 1981, Pylyshyn 1984; for (neo)-connectionism see Rumelhart & McClelland 1986, Smolensky 1988, Amit 1989; for comparative studies, see Clark 1989, Andler 1990 (mostly sympathetic to connectionism), and Fodor & Pylyshyn 1988 and the two other papers in Pinker & Mehler 1988 (mostly hostile to connectionism).

3. Jean Mosconi's unpublished doctoral thesis (Université Paris I, 1988) is certainly the place to start. I deeply regret not having had a chance to read it before setting on this investigation.

4. Depending on whether it does or doesn't have circles, and is or isn't outfitted with a tape and the functional equivalent of a read-write head.

5. Quoted by S. Papert, in McCulloch 1965/1988, p. xxviii.

6. This point — the non-intrinsic character of the embodiment, by some node in a net, of any particular proposition or formula — is made by Fodor 1986 (and again in Fodor & Pylyshyn 1988), but is meant by him to apply (with supposedly devastating effects) to the nets of *neo*-connectionism. This struck me, when I first came upon it, as a complete misunderstanding of *that* kind of connectionism, and led me to consider the problem of locating the difference which would explain why a perfectly cogent argument against nets as construed by McCulloch and Pitts appears out of place in the context of contemporary neural net research.

7. McCulloch 1965/1988, p. 21. The idea expressed in this quote is explicitly attributed to "one of [the authors]", *viz.* of course McCulloch (in McCulloch 1961, he tells the story of his discovery: "In 1929 it dawned on me ...". McCulloch 1965/1988, pp. 8-9). McCulloch was born in 1898, Pitts in 1923. In the conclusion of the paper (p. 38) it is again asserted that "in psychology (...), the fundamental relations are those of two valued logic.".

8. On the idea, crucial to modern science in general (and not just cognitive science), of the common 'discipline' imposed by mathematics upon symbols and by nature upon the things symbolized, see Cummins 1989, pp. 27-29.

9. This, I presume, is the fault that Heinz von Foerster, the leader of the 'second cybernetics', who founded and headed the Biological Computer Laboratory at the University of Illinois in the 1960's, finds in McCulloch's concept of information:

he confused it, or so claims von Foerster, with a mere signal. See Lévy 1985, Livet 1985 and this volume.

10. McCulloch 1965/1988, p. viii. It should be noted however that the 1943 paper contains no reference to Lloyd's work.

11. That *is* crudely put, for, as is well known, contemporary theories of vision actually favour a largely symbolic conception of vision. As Marr puts it, "the critical point seems to be that even very early vision is a highly symbolic activity" (*o.c.*, p. 350). But this, if true, is by no means an *analytic* truth; it is a hard-won empirical conjecture.

12. Rumelhart & McClelland 1986. The relative numerical importance of PDP makes it tempting to call the whole 'input/output' movement after its main component, but for a number of reasons I wish to resist the temptation.

13. As pointed out, in particular, by Minsky and Papert in their new fore- and afterwords to their classic 1969 treatise on perceptrons.

14. As stressed, among others, by Fodor & Pylyshyn 1988.

15. See Fodor & Pylyshyn 1988, Smolensky 1988a, Visetti 1990, Andler 1990, 1990a.

16. The most thorough defense of this approach is Smolensky 1988.

17. The key reference here is Amit 1989.

References

Amit, D.J., 1989, *Modeling Brain Function. The world of attractor neural networks*, Cambridge U.P., Cambridge.

Anderson, J.A; Rosenfeld, E., 1988 (A&R), *Neurocomputing. Foundations of Research*, MIT Press, Cambridge, Massachusetts.

Andler, D., 1986, Studying cognition today, report to the European Science Foundation, *Eidos*, vol. 5, pp. 177-225; with two appendices in *Artificial Intelligence and Language*, H. Sinding-Larsen (ed.), Oslo, Tano, pp. 201-246.

Andler, D., 1987, Progrès en situation d'incertitude, *Le Débat*, vol. 47, pp. 213-234.

Andler, D., 1990, Connexionnisme et cognition: à la recherche des bonnes questions, *Revue de Synthèse*, série générale tome CXI, vol. 4, nos. 1-2, pp. 95-127.

Andler, D., 1990a, Structures, espaces et temps dans les réseaux connexionnistes, *Intellectica*, nos. 9-10, pp. 283-297.

Boolos, G.S.; Jeffrey, R.C., 1974, *Computability and Logic*, Cambridge U.P. Cambridge.

Clark, A., 1989, *Microcognition. Philosophy, Cognitive Science, and Parallel Distributed Processing*, MIT Press, A Bradford Book, Cambridge, Massachusetts.

Cummins, R., 1989, *Meaning and Mental Representation*, MIT Press, A Bradford Book, Cambridge, Massachusetts.

Dupuy, J.-P., 1985, L'essor de la première cybernétique (1943-1953), *Cahiers du CREA*, no. 7, Ecole Polytechnique, Paris, pp. 7-140.

Feldman, J.A., Ballard, D.H., 1982, Connectionist models and their properties, *Cognitive Science*, no. 6, pp. 205-254; repr. in A&R.

Fodor, J.A., 1975, *The Language of Thought*, Thos. Crowell, New York; repr. Harvard U.P., Cambridge, Mass.

Fodor, J.A., 1981, *Representations: Philosophical Essays on the Foundations of Cognitive Science*, MIT Press, A Bradford Book, Cambridge, Massachusetts.

Fodor, J.A., 1983, *The Modularity of Mind*, MIT Press, A Bradford Book, Cambridge, Massachusetts.

Fodor, J.A., 1985, Fodor's guide to mental representation: The intelligent Auntie's vade-mecum, *Mind*, vol. XCIV, pp. 76-100.

Fodor, J.A., 1986, Information and association, *Notre-Dame Journal of Formal Logic*, vol. 27, pp. 307-323.

Fodor, J.A., 1987, *Psychosemantics. The Problem of Meaning in the Philosophy of Mind*, MIT Press, A Bradford Book, Cambridge, Massachusetts.

Fodor, J.A.; Pylyshyn, Z., 1988, Connectionism and cognitive architecture: A critical analysis, *Cognition*, vol. 28, nos. 1-2, pp. 3-71.

Gardner, H., 1985/1987, *The Mind's New Science: A History of the Cognitive Revolution*, Basic Books, New York.

Goldstine, H.H., 1972, *The Computer from Pascal to Von Neumann*, Princeton U.P., Princeton.

Graubard, S.R. (ed.), 1988, *The Artificial Intelligence Debate. False Starts, Real Foundations*, MIT Press, A Bradford Book, Cambridge, Massachusetts; and special issue of *Daedalus*, vol. 117, no. 1, Winter 1988.

Grossberg, S., 1980, How does the brain build a cognitive code ?, *Psychological Review*, vol. 87, pp. 1-51; repr. in A&R.

Haugeland, J. (ed.), 1981, *Mind Design*, MIT Press, A Bradford Book, Cambridge, Massachusetts.

Haugeland, J., 1978, The nature and plausibility of cognitivism, *Behavioral and Brain Sciences*, vol. 1, pp. 215-226; repr. in Haugeland 1981.

Hebb, D.O., 1949, *The Organization of Behavior*, Wiley, New York.

Hebb, D.O., 1980, *Essay on Mind*, Lawrence Erlbaum Associates, Hillsdale, NJ.

Hinton, G.E.; Anderson, J.A., 1981, *Parallel Models of Associative Memory*, Lawrence Erlbaum Associates, Hillsdale, NJ.

Hopfield, J., 1982, Neural networks and physical systems with emergent collective computational abilities, *Proceedings of the National Academy of Sciences*, vol. 79, pp. 2554-2558; repr. in A&R.

Lévy, P., 1985, Analyse du contenu des travaux du Biological Computer Laboratory, *Cahiers du CREA*, no. 8, Ecole Polytechnique, Paris, pp. 155-191.

Livet, P., 1985, Cybernétique, auto-organisation et néo-connexionnisme, *Cahiers du CREA*, no. 8, Ecole Polytechnique, Paris, pp. 105-153.

McClelland, J.L.; Rumelhart, D.E. and the PDP Research Group, 1986, *Parallel Distributed Processing: Explorations in the Microstructure of Cognition. Vol. 2: Psychological and Biological Models*, MIT Press, A Bradford Book, Cambridge, Massachusetts.

McCulloch, W.S., 1965/1988, *Embodiments of Mind*, MIT Press, Cambridge, Massachusetts.

McCulloch, W.S., 1961, What is a number, that a man may know it, and a man, that he may know a number ?, *General Semantics Bulletin*, nos. 26-27, pp. 7-18; repr. in McCulloch 1965/1988, pp. 1-18.

Minsky, M.; Papert, S., 1969/1989, *Perceptrons*, MIT Press, Cambridge, Massachusetts.

Newell, A., 1983, Intellectual issues in the history of artificial intelligence, in *The Study of Information: Interdisciplinary Messages*, F. Machlup and U. Mansfield (eds.), Wiley, New York.

Newell, A.; Simon, H.A., 1976, Computer science as empirical enquiry: Symbols and search, *Comm. Am. Ass. Computing Machinery*, vol. 19, pp. 113-126; repr. in Haugeland 1981.

Petitot, J., 1990, Le Physique, le Morphologique, le Symbolique: remarques sur la vision, *Revue de Synthèse*, série générale tome CXI, vol. 4, nos. 1-2, pp. 139-183.

Pinker, S.; Mehler, J., 1988, *Connections and Symbols*, MIT Press, A Bradford Book, Cambridge, Massachusetts.

Pitts, W.; McCulloch, W.S., 1947, How we know universals: The perception of auditory and visual forms, *Bulletin of Mathematical Biophysics*, vol. 9, pp. 127-147; repr. in McCulloch 1965/1988 and in A&R.

Pylyshyn, Z., 1984, *Computation and Cognition. Toward a Foundation for Cognitive Science*, MIT Press, A Bradford Book, Cambridge, Massachusetts.

Rosenblatt, F., 1958, The perceptron: A probabilistic model for information storage and organization in the brain, *Psychological Review*, vol. 65, pp. 386-408; repr. in A&R.

Rumelhart, D.E.; McClelland, J.L. and the PDP Research Group, 1986, *Parallel Distributed Processing: Explorations in the Microstructure of Cognition. Vol. 1: Foundations*, MIT Press, A Bradford Book, Cambridge, Massachusetts.

Shannon, C.E., 1938, A symbolic analysis of relay and switching circuits, *Trans. AIEE*, vol. 57, pp. 713 sq.

Smolensky, P., 1988, On the proper treatment of connectionism, *Behavioral and Brain Sciences*, vol. 11, pp. 1-37.

Smolensky, Paul, 1988a, Connectionism, constituency, and the language of thought, in *Jerry Fodor and His Critics*, B. Loewer and G. Rey (eds.), forthcoming.

Varela, F.J, 1979/1989, *Principles of Biological Autonomy*, Elsevier/NorthHolland, New York; New French edition, Autonomie et connaissance, Le Seuil, Paris.

Visetti, J.-M., 1990, Modèles connexionnistes et représentations structurées, *Intellectica*, nos. 9-10, pp. 167-212.

SECOND CYBERNETICS: A DOUBLE STRATEGY FOR REPRESENTING COGNITION*

Pierre Livet

Centre de Recherche en Epistémologie Appliquée, CREA
Ecole Polytechnique & CNRS, Paris

Abstract

The fact that the programme of second order cybernetics (which was called by Von Foerster the theory of the observing systems), failed to come up to expectations on the one hand, and the simultaneous success of Artificial Intelligence (AI) on the other hand, are our starting points to inquire into the problem of the modelling of cognition. Why didn't cybernetics consider AI as a prolongation of itself ? Why did Von Foerster see first cybernetics as inadequate to model cognition ? Why have Von Foerster's tessellations of 'cognitive tiles' passed into oblivion, whereas connectionist models are considered to be serious propositions to get at a model of cognitive abilities ? On the basis of Von Foerster's criticism of first cybernetics and his interpretation of memory, we inquire into the particular view on cognition in cybernetics of second order. The concept of 'inverted reductionism' is meant to give expression to the specificity and the inadequacies of this view.

Daniel Andler formulated for us the relations between paleo-connectionism and neo-connectionism. Paleo-connectionism, however, was in the end nothing but cybernetics. Six years ago, Jean-Pierre Dupuy, Isabelle Stengers, Pierre Lévy and I carried out a research about cybernetics and its history, in contradistinction to Artificial Intelligence, and as an anticipation of the theories of self-organization and of the recent theory of networks of automata (*Cahiers du CREA*, 1985). In this history, we find three moments: first cybernetics, second cybernetics and connectionism. I will concentrate myself on the second cybernetics, and especially on its leader Von Foerster. Cybernetics is the science of systems which regulate and steer themselves. Wiener, McCulloch and Ashby, using Shannon's notion of information and the notion of feed-back, built homeostatic systems. But Von Foerster thought these notions were not sufficient in order to build a theory of

G. Van de Vijver (ed.), New Perspectives on Cybernetics, 147–156.

cognitive systems. Cybernetics was necessary to understand observed systems, but as cognitive systems are observing systems, a theory of observing systems had to be developed, which he called second cybernetics.

Actually, second cybernetics was rather a failure. But failures are sometimes more interesting than successes, at least for the philosophy of sciences. We can remind here that this failure was contemporaneous with a comparative success, the success of Artificial Intelligence. Second cybernetics was one way to model cognition; Artifical Intelligence was another way, but they were contemporaneous. Why did the latter progress whereas the former stopped ? After all, the first cybernetics seemed to be able to produce interesting results in the field of the simulation of cognition. Daniel Andler has shown that McCulloch in his first cybernetics-approach had already thought of his network of formal neurons as a machine which operated on symbols. So the first cybernetics was able to cope with the problem of cognition, as settled in the terms of AI: as a computation upon sequences of symbols (eventually provided with semantics). Why didn't the cyberneticians think of AI as a continuation of their own work ? Why did Von Foerster think that first cybernetics was unable to cope with cognition, i.e. with observing systems ? Why do the connectionist successors of McCulloch's network of formal neurons, the automata networks, have the capacity of modelling some parts of cognition, while Von Foerster's tessellations of 'cognitive tiles' don't ? I will try to give a sketch of some answers to these questions.

First, let us return to the limitations of the first cybernetics and compare the criticisms of Von Foerster with our present critical judgment (we have of course the unmerited advantage of knowing the future of cybernetics, the development of AI, and the present expansion of the research about automata networks). Von Foerster made two main objections to Wiener and the first cybernetics. First, they reduce information to signal. Secondly, they reduce learning to homeostasy and reflexivity to feed-back (and, correlatively, it would be better to say homeodynamics than homeostasy). Von Foerster made no objections to the McCulloch networks. But, as he proposes his cognitive tiles, we can see that they are much more complex elements than the formal neurons of McCulloch. Von Foerster's 'non-trivial machines' involve functions of functions, which makes them second order entities and not only Boolean operators.

From the new outlook we have got from automata networks and AI, we can make two criticisms against McCulloch's networks. On the one hand, McCulloch reduced the symbols his machine was supposed to compute to numerical symbols or abstract symbols. He didn't think of his machine as manipulating general symbols, such as sentences. Hence, he satisfied himself with the logical possibility of computing the basic logical functions. He didn't try to build concrete machines which computed real cognitive problems. As a matter of fact, at his time it was not possible to compute and to simulate massively parallel computations. On the other hand, he saw his network as a way to demonstrate the identity between a neural system and a Turing machine. He didn't really see the networks as a way to a new style of computation or as means of producing new phenomena, such as collective phenomena, stable states, attractors, and so on (yet he defined the heterarchical property of a network in opposition to properties of trees). Although he studied some features of networks, he didn't explore systematically the class of functions some architectures of networks are able to compute, and he didn't define specific tasks for which networks are particularly useful. But of course, second cybernetics didn't make any progress in this direction either.

Let us go back to the first cybernetics. Are the criticisms we can make against it, with the advantage of the historian's distance, the same as those of Von Foerster, who criticizes the reduction of information to signal. Perhaps he is right. But I think that first and second cybernetics both made the same mistake, or more precisely, adopted the same wrong strategy with regard to this notion of information. Wiener and Von Foerster (and especially Ashby, the other great figure of the second cybernetics) used it as an invariant magnitude, which remains constant through its successive transformations. This point of view was quite natural for Von Foerster, being a physicist. As a matter of fact, nothing in itself is wrong with this point of view. But in so far as they tried to model cognition however, I think it was the wrong strategy. For the problem of cognition is not finding an invariant quantity, but finding the adequate transformations which enable our cognitive processes to represent new relations, to modify the format of our representations, to build new rules, and so on. The focusing on the notion of information in first and second order cybernetics, as opposed to the strategy of AI of not focusing on information but on information processing, is very instructive indeed. Cyberneticians are assuming a substratum for forms, which is the information which encodes all forms. AI is concerned with the modification of forms and with rules for passing from some sequences of symbols to other ones (the information is supposed to be transmitted in the machine, but this is in fact the informatician's problem).

Rather surprisingly, second cyberneticians, who were interested in cognition and not only in regulated systems, seemed to be indifferent to the problem of how to assign good categorizations and how to recognize specific forms. Their approach was abstract and general, and they brought into the field of cognition the equivalent of the focalization upon invariant 'information', which was useful when the problem was the transmission of information for realizing regulation.

The most striking example of this trend is the so-called demonstration of the impossibility and self-contradictory character of self-organization, proposed by Ashby. For him, 'organizing' means changing from a bad organization to a good one; 'self-organizing' means changing automatically from a bad organization to a good one (Ashby, 1962, p. 267). Hence, we still have to wait for a definition of what it is to be considered as a good or a bad organization. For instance, a good organization would be one which provides stable classifications of the signals sent by the environment, useful categorizations of the different inputs, and classes of outputs which fit both the environment and the system. But Ashby only states that "organization is relative; and an organization that is good in one context may be bad in another" (*Ibid.*, p. 263), and then his demonstration follows. If we remember that a machine is a function, 'self-organization' requires that this function f, which determines the changes of the internal state of the machine, will itself change in function of these states (if it is a bad state, it changes into a good one). If this would be the case, we would define a meta-function, whereby our previous function would be a variable which changes as a function of time. Thus, Ashby confronts us with a dilemma. Either self-organization is impossible (as it implies that we change the function when its variable changes, which is the negation of the distinction between a function and a variable), or it is trivial. As a matter of fact, this dilemma is a trilemma, as self-organization can be trivialized in two ways. The first way consists in defining a hierarchy of meta-functions; the second way is a variant of the first one. If we are convinced by Ashby's demonstration of impossibility, we nevertheless may think of organization, and of changing this organization into a good one, as a system going towards an equilibrium. But the system has now become the conjunction of an organization and its environment (so that the whole system is an isolated one), and its dynamics are very complex. Going from any state to an equilibrium is going from a larger number of states to a smaller one. If the remaining selected states are sufficiently numerous however, the organization can switch from one equilibrium to another, and hence the equivalent of self-organization is possible. But this is trivial, as the organization in this case is a degenerated case of a dynamic system.

However, I think the whole demonstration misses the point. Organization and self-organization are not only equilibria, but "good" equilibria or sets of good equilibria. Let us look at the automata networks. Any dynamics of these networks may lead us to attractors or stable states. The problem is to avoid "spurious states" (too small and also irregular attractors, that is), as well as huge attractors, which would classify everything in the same basket. Therefore, the problem is to define and to produce not only "good" attractors, but also, if possible, attractors attuned to our own categorizations (not the same categorizations but compatible with them). Hence, self-organization *means* going from one equilibrium to another. The important thing is not this triviality, but the categorization of equilibria. Thus, by putting the problem in an abstract and general fashion, Ashby missed the whole point.

Nevertheless, the ambition of the Biological Computer Laboratory (B.C.L.) (Von Foerster's and Ashby's laboratory) was to model cognition, even the higher degrees of cognition such as reflexivity. In order to achieve this project, with the limitations of their abstract and rather trivial starting point, they adopted a double strategy. On the one hand, Von Foerster thought it was necessary, in order to cope with high cognitive functions, to climb in the hierarchy of logical levels, to hasten into an epistemologic climbing. But, on the other hand, he tried to use at each level the simplest mechanisms. For instance, he thought that the storage of information was not necessary to model memory. A finite state machine was sufficient, because its present internal state depends on the previous internal state and the previous input, and so on and so forth by recursion.

I propose to call this double strategy an *inverted reductionism*. It is a reductionist strategy because complex cognitive functions are reduced to simpler ones (memory is reduced to dependence on the previous states). This reductionism is inverted, as these simpler functions are lifted up to a higher logical level (variables become functions, functions become functionals). This strategy was a direct consequence of Ashby's demonstration of impossibility, in which self-organization was either impossible if it means taking the function as changing with the variable, or trivial if it means taking the function as a variable of time and in so doing defining a functional. Von Foerster opted for triviality and compensated it by climbing up through the logical levels. The two characteristics of this strategy collapse into one with the notion of reflexivity; for reflexivity may be the loop by which a rather simple function re-enters under the same function and is lifted up to a higher logical level. In reflexivity, the complexity is not given by the function itself (although this is possible), but by the recursivity by which the function re-enters under itself.

Here again, the contrast with the automata networks may be illuminating. The complexity of the collective phenomena of these networks is not the result of climbing in the logical hierarchy, but is the result of the collective interactions between automata, of the loops of iterations by which these interactions are computed, and of the plurality of structural levels (and not logical ones). Collective interactions have taken the place of logical climbing, structural levels the place of logical ones and loops of iteration the place of reflexivity.

The three more famous among Von Foerster's texts can be read as examples of this strategy. In 'On Constructing a Reality' (1973), Von Foerster progressively reduces the exteriority of reality to a reflexive or recursive loop. A cognitive process is the computation of a reality. But since computation deals with descriptions and not with objects, cognition is computation of the description of a reality. Cognition of a higher order is not working on descriptions, but on descriptions of descriptions. Computation of descriptions is merely computation, so cognition is computation recursively applied to itself (reflexivity is not so perfect here, as Von Foerster said, as cognition of the lowest order remains a description of something; reflexivity is reduced here to recursivity). Nothing is wrong in this way of thinking, but how can it be useful ? Recursivity is evidently a characteristic of cognition, but we still have to produce the specific recursive functions which model cognition and to compute the results of their recursivity.

So let us go back to some papers which are a bit less speculative and perhaps more technical, such as 'Molecular Ethology' (1970), or 'What is a memory that it may have hindsight and foresight as well ?' (1969). Here, Von Foerster tries to demonstrate that the formalism of a finite state machine can be used to model 'non-trivial machines'. He tries to reformulate cognitive capacities, like memory, in such a way that the finite state machine formalism may model these capabilities and he proposes some combinations of elementary cognitive processors, called 'cognitive tiles'.

A trivial machine is a function $y = f(x)$. Given the input x and the function f, the output y is given. A finite state machine is a function $y = f(x,z)$, where z is the present internal state of the machine. If we don't know this internal state, we cannot predict the output y. Z can be defined recursively as being dependent on previous y and previous z. Previous z may be finally a function of previous x and of the initial y. So, finite state machines are non-trivial machines, in which the present output may be dependent on the sequence of inputs through the past history of the system and on some remote output.

Therefore, we may reformulate once more the $y = f(x,z)$ and say that it is equivalent to $y = h_z(x)$, were h_z is a function which changes with each z. Thus, the non-trivial machine is a sequence of different trivial machines (the functions of which are the successive $y = h(x)$). Here again, we find the influence of Ashby's conception. But all that is rather trivial. Of course, if a finite state machine has sufficiently numerous states, it can exhibit a very complex behaviour. But we either have to give it all those states, or we have to define all the different functions h and construct all the trivial machines. There is nothing to be gained in that way. Hence, Von Foerster uses his double strategy. First, he reduces memory to dependence on previous history. This is neither a real nor a successful reduction: Von Foerster is well aware of the fact that finite state machines don't have memory, because they cannot recapture any one of the past states or outputs (Von Foerster, 1970, p. 222). He concludes: "we have restored the original meaning of concepts like memory as various manifestations of a more inclusive phenomenon, namely, cognition" (*Ibid.*, p. 247). This 'original' meaning is compatible with non-trivial machines (that is, finite state machines). How can an unsuccessful reduction be seen as a reintegration of memory in the more global and complex phenomena of cognition ? By using the second branch of the strategy, climbing up the logical levels. Von Foerster imagines a finite state machine to be transformed into a finite function machine. This machine will not only change its states, but also its programmes. For this machine, the reference to past events is not only to its inputs, but also to its history of potential actions. If we can define 'eigenfunctions', which satisfy the equation defining the recursive dependence of the behaviour of the machine on its previous inputs and on its previous output functions, we may save the history of potential functions. These 'eigenfunctions' will give us the "present experience of an experience in the past" (*Ibid.*, p. 236). Climbing up the scale of functions to functionals therefore makes us able to compensate for the reduction of memory through an integration of the past history in the present internal state of a finite state machine, which becomes effectively non-trivial. But, as Von Foerster honestly admitted, "this formalism does not specify any mechanism capable of performing the required computations" (*Ibid.*, p. 236). As a matter of fact, Von Foerster's double strategy jumps over the problem of memory; with his non-trivial machine, we have either the experience in the past (when it was present), or the present experience of the experience in the past, but not the storage and the recalling of the experience as a past experience. Everything is present in the finite state machine. Von Foerster doesn't solve the conceptual problem nor the practical one: constructing a machine which has a memory, using it as a cognitive memory (a machine which learns).

However, with his tessellations of cognitive tiles, Von Foerster seems to come very near to the actual connectionism. It will be instructive to compare those tessellations and automata networks.

A cognitive tile is an aggregate of simple processors, disposed so as to form a cross. To its left, it has a sensory receptor, which interprets the inputs of the environment by computing relations between observed activities (external and internal) with respect to the 'self'. To its right, the output is sent through an element T which translates it into a 'universal internal' language, accessible to the other cognitive tiles of the system. At the top of the cross, we find a delay loop, which operates on the feed-back channel coming from the output. At the bottom of the cross, another processor, which operates on another similar feed-back channel, computes all the relational information of the system's own actions as far as they refer to the 'self' (so it computes the relations of the system's actions to its goals). Finally, at the centre of the cross, we have a finite function machine, of which the different functions are changing according to the eigen-states of the machine (Von Foerster, 1969, p. 46).

Here again, the recursivity of the central machine and the two feed-back loops mask the problem: how is the set of functions which enables the machine to start, defined ? How are the 'relations with respect to the self' defined ? Is the 'self' the central finite function machine or the whole aggregate ? How are the system's actions related to its 'self' ? Are the modalities of assigning weights to these outputs, in accordance with the 'goals and desires' of the system, defined once and for all externally, or are they recursively modified ? Von Foerster only tells us that with just a few cells making up such a computer (the central machine, the delay loop and the goal loop), the number of different solutions is for all practical purposes infinite. He thinks that establishing a detailed account of how the inhibitions and facilitations of a neural network could realize such a computer, is of no importance (*Ibid.*, p. 45).

That was the problem the automata networks were coping with. If we try the comparison, we find several differences. First, the automata of a network are very simple operators (threshold functions or Boolean operators). The cognitive tile was already an aggregate of processors, having an internal structure. The elementary automaton of the network has only the material structure necessary to simulate its function. It is not a functional, so it needn't be called a 'self' at all. The recursivity doesn't take place in just one single automaton, but in the whole network by the iteration loops and the closure of the connections.

In this case, a translator into a universal internal language is not necessary: the automaton simply sends its states as outputs to its neighbours. The important thing is the architecture of the connections and the algorithm by which we can compute the weights to be assigned to each connection for the network to be able to learn a specific form. In Von Foerster's cognitive tiles, these weights were inserted in the goal loop, which means they were internal to one tile. In the automata network, they are external to the automata and define the behaviour of the network. However, Von Foerster imagined some rules to connect his tiles with each other. For instance, the goal processor of one tile can be connected with the delay processor of one of the right tiles, and, of course, the output translator of one of the left tiles can be connected with the input processor of one of the right tiles. The delay loop of one left tile can be connected, I think, with the input processor of one right tile. But this goal processor cannot be connected with the output translator of a left tile (so no loop between tiles is permitted). Von Foerster thought of his tessellations more as a combination of computational elements than as a collective interactive system. Each tile has a cognitive result and the result of tessellation will be a macro-result. The computation of each automaton, at least at the lowest level, doesn't carry any cognitive meaning in itself though the cognitive result is the result of the whole network.

Finally, as we have already stated, the notion of level doesn't play the same role in cognitive tiles as in automata networks. In cognitive tiles, the difference between levels is a logical one, which takes place inside the cognitive tile: it is the difference between the functional and the previous inputs and the previous output functions, which makes the 'non-trivial machine' dependent on its past history and its modalities of experience (that is, its 'eigen-states'). This dependence is materialized by the loops of the delay channel and the goal channel. In automata networks, the difference of levels is a structural and a functional one. It takes place as an intermediary between the elementary automata and the collective network. So recursivity (iteration, in fact) is not a characteristic of elementary automata, but of the architecture of the network, and of the rules of its operations. Of course, one of the main reasons for that difference was that Von Foerster had no means to simulate a parallel interaction on a computer, nor to observe the emergence of collective states. Therefore, he concentrated on the basic features of an elementary tile and thought of their collective operating merely as a combination of the elementary operations. That's the reason why, I think, he describes the features of his cognitive tiles in an abstract and general fashion, and why he doesn't pay attention to the building of an effective collective structure and to the peculiar properties of an assembly of cells (tiles or automata), as compared to the properties of its elements. Second cybernetics only dealt with the formal general features which a cognitive element must exhibit. In my opinion, with the advantage of knowing the continuation of the story, it was an erroneous strategy. It is better to study first the properties of a concrete and sufficiently complex model, even if it may be called 'cognitive' only by analogy, and to confront it with cognitive achievements. If automata networks don't realize our cognitive functions precisely, they at least make us able to discover new forms of cognitive operations and to study new formal (but implementable) objects.

Hence, what is the moral of this story ? Inverted reductionism is not a good strategy if we want to construct a formal representation of cognitive activities. Inverted reductionism is raising up the logical level of rather trivial operations, in order to make it possible to simulate complex cognitive properties. If we insist on being reductionists and on following the way opened by the paleo-connectionists and first cyberneticians, we are better off to use the capacities of collective assemblies of trivial operators and to build structural levels, than to climb up logical levels. The latter is more difficult, as we have to find the better algorithms, the more attractive architecture and the laws of such collective systems. After all this, the confrontation with real cognitive tasks may still be deceptive. The strategy of using a roundabout in an effective way is better than the strategy of using a short-cut in a logico-speculative way.

* This article was first published in *Communication & Cognition*, vol. 23, nos. 2-3.

References

Ashby, R.W., 1962, Principles of the self-organizing system, in *International tracts in computer science and technology and their application*, vol. 9: Principles of self-organization, H. Von Foerster and G.W. Zopf (eds.), Pergamon Press, Oxford, pp. 255-278.

Cahiers du CREA, 1985, no. 7: Histoires de cybernétique, Paris.

Von Foerster, H., 1973, On constructing a reality, in *Environmental Design Research*, vol. 2, W. Preiser (ed.), pp. 35-46.

Von Foerster, H., 1970, Molecular Ethology: An Immodest Proposal for Semantic Clarification, in *Molecular Mechanisms in Memory and Learning*, G. Ungar (ed.), Plenum Press, New York, pp. 213-249.

Von Foerster, H., 1969, What is Memory that it May Have Hindsight and Foresight as well ?, in *The Future of the Brain Sciences*, Proceedings of a Conference held at the New York Academy of Medecine, S. Bogoch (ed.), Plenum Press, New York, pp. 19-64.

THE BRINGING FORTH OF DIALOGUE: LATOUR VERSUS MATURANA

H. Timon Meynen

University of Utrecht

"The notion of system is of no use to us, for a system is the end product of tinkering and not its point of departure (...). For a system to exist, entities must be clearly defined, whereas in practice this is never the case; functions must be clear, whereas most actors are uncertain whether they want to command or obey; the exchange of equivalents between entities or subsystems must be agreed, whereas everywhere there are disputes about the rate and direction of exchange. Systems do not exist, but systematizing is common enough; everywhere there are forces that oblige others to play the way they have always played (...)." (Latour, 1984/1988, p. 198)

Abstract

After introducing the problem and a short impression of the work of Humberto Maturana, I elaborate on his ideas concerning science and the scientific explanation (1.) and objectivity (2.). In an evaluation of these views I claim that we cannot accept some consequences of Maturana's theory, because they leave no space for discussion (3.). I then turn to Bruno Latour, and after a short introduction (4.) proceed to claim that there are some interesting similarities between Latour and Maturana (5.), especially in their respective ideas on the praxis of explaining. I further claim that Latour's idea of 'infrareflexivity' constitutes a way out of the problems with Maturana's theory (6.). I conclude with a 'plea for taking dialogue seriously' (7.). Along the way, it turns out, I could show Latour a counterexample of his claim that systems theorists are not aware of their own role in creating 'systems'.

G. Van de Vijver (ed.), New Perspectives on Cybernetics, 157–174.

Introduction[1]

The citation above stems from an important author in recent relativistic 'science of science', Bruno Latour. His view on cybernetics applies to a lot of modern use of systems theory in the social sciences, including (cognitive) psychology. But there are some exceptions also. With respect to the *process* of *'tinkering'* which Latour mentions, we are talking about the fact that, when claiming that for instance human beings are (observed) systems, one has to admit that they are at least *observing systems* too.

Since the seventies this has become a theme in cybernetics, resulting one could say in 'second-order cybernetics'. The system theorist Heinz von Foerster once appointed Humberto Maturana as the main thinker on this new territory (Von Foerster, 1981). In this paper, I shall investigate whether Maturana can escape the criticism of Latour. My claim will be that he can, provided that he listens to Latour. Likewise Latour should listen to Maturana, in order to find a counterexample for his picture of the users of words like 'system'. Finally, I shall claim that this way of dealing with each other could function as some sort of 'foundation' for theory and philosophy of science, and for the (social) sciences.

If one intends to give a summary of a theory like Maturana's, one is confronted with the difficulty that a summary already presupposes a kind of shared 'framework' between reader and writer. Whereas Maturana's theory itself precisely aims at creating a whole new framework, within which it is possible to explain how such a shared framework comes or has come into existence. Although Maturana uses words which are adopted from a common language, this handhold disappears slowly once it becomes clear that Maturana defines all the words he uses only with respect to each other (or with respect to earlier work of his own).

Maturana started his academic life with some experiments in the perception of frogs.[2] As a result of his findings, he changed thinking about perception by claiming that we-as-observers no longer could and should maintain, that we are able to determine whether there really exists a difference or correspondence between the perception of a living system (nowadays called a 'representation') and the actual environment. During the seventies Maturana elaborated - partly in cooperation with Francisco Varela - this, what he called 'autonomy' of living systems, for biology and cybernetics (Maturana & Varela, 1980). For too long, biology had confined itself to the study of the development and reproduction (i.e. evolution) of living systems. Now, Maturana and Varela sought a way to talk scientifically about living systems as a living unity, in a particular place, at a particular time.

Furthermore, Maturana and Varela explicitly wanted to take into account that people, and so the authors themselves, are living systems too. This led Maturana in the eighties to theorizing about specifically human phenomena like language, thought, (self-)consciousness, science, love, society, etc. As an example of Maturana's style of thinking and at the same time as a starting point for a 'critique' we take his theory of 'cognition', and focus on his idea of 'science' and the 'scientific explanation'.

1. Maturana on science and cognition

According to himself the aim of the article called 'The Biological Foundations of Self-Consciousness and the Physical Domain of Existence' (Maturana, 1987) is the following:

> "My purpose in this essay is to explain cognition as a biological phenomenon, and to show, in the process, how language arises and gives origin to self-consciousness, revealing the ontological foundations of the physical domain of existence as a limiting cognitive domain." (*Ibid.*, p. 324)

Maturana generally opens as reflexive as possible, and this time he does so by introducing "two unavoidable experiential conditions" (*Ibid.*, p. 324):

1) "cognition is a biological phenomenon" (*Ibid.*, p. 324). This we know because every change in our 'biology' leads to a change in our 'cognition'; and

2) "we, as is apparent in this very same essay, exist as human beings in language using language for our explanations" (*Ibid.*, p. 324). This means that every change in our cognition (use of language) also changes our nervous system, simply because we describe it differently, see it differently, for example as a telephone-exchange instead of a walnut.

These two statements thus refer to the phenomena-to-be-explained, which are dealt with in this text, as well as to the 'unavoidable instruments' of explanation: that what we must presuppose to be at work in the process of explanation. The goal of the rest of Maturana's essay is to make everything that is said about cognition-as-subject compatible with the fact that it is also a condition for saying anything.

Maturana is a biologist, and as such he is also a scientist, so he will give us a scientific explanation of cognition as a biological phenomenon. Maturana finds himself, at the very moment he calls himself a biologist and a scientist - and thus human "... here and now in the praxis of living, in the happening of being human, in language languaging, in an *a priori* experiental situation in which everything, that is, everything that happens, is and happens in us as part of our praxis of living." (Maturana, 1987, p. 326).

If this is the case and we want to talk about what happens in this 'praxis', then this is a retrospective comment on what happened and it can never take the place or change the form of that what actually happened. An explanation always takes place in a 'meta-domain' with respect to the things to be explained. Furthermore, when we give such an explanation about something in our praxis of living, then it is always an answer to a question about the *origin* of some given phenomenon. The listener rejects or accepts this explanation on the basis of some criterion or another; the 'modern' scientist employs 'the scientific explanation' as a criterion for the acceptability (validity) of statements.

According to Maturana, scientists do not usually think about the conditions of their labour. That is why Maturana can pass by that huge amount of philosophies, methodologies and theories of science, which a theoretical psychologist for instance has to stow away in his or her daily life. And Maturana can simply give an 'operational (and so an experiental) specification' of what constitutes a scientific explanation as a criterion of validation of propositions. This specification can at the same time be considered as a reformulation of the conditions which have to be met for the community of 'standard observers' in order to accept a particular mechanism as a scientific explanation:

1. The phenomenon has to be described by naming what has to be done to witness the phenomenon.

2. The proposal of a 'generative mechanism' (often called the explaining hypothesis) which brings forth the phenomenon in 1. We have already seen that Maturana claims that 2. can never take the place of 1.

3. The inference of other phenomena which a standard observer should have to witness as a result of the operation of the same mechanism, including the things an observer has to do to actually see them.

4. The witnessing of the phenomena mentioned in 3.

Such a reformulation of the praxis of living of the standard observer, the Unknown Warrior of modern science, enables Maturana to infer some important aspects of scientific statements. In the first place, it follows from the description that scientific statements are '*consensual statements*' and are only valid within the community of standard observers that produced them. So it is the conversation between scientists which precedes all. This means that science is not a way to reveal an independent reality. As a result the '*universality*' which is often said to be a feature of scientific statements does not depend upon such a reality but rather on the universality of the community of standard observers: it must be possible for both Russians and Americans to be a member of this community, as well as for Chileans, Dutch and French.

Secondly, it is often said of scientific explanations that they are *reductionistic*. According to Maturana this is not the case because when explaining, we are not expressing phenomena of the first domain (the 'praxis of living') in phenomena of the second (the domain of scientific explanations). The scientific explanation and its generative mechanism take place in another domain than that of the phenomena-to-be-explained. The generative mechanism is produced *ad hoc*, accepted or rejected *a priori*, and in principal doesn't need any justification (Maturana, 1987, p. 330). The domain of scientific statements thus constitutes a domain of 'operational coherences' in the praxis of living of the observer. It is notable that a lot of things that, to many psychologists, guarantee the scientific character of their work - measurement, quantification and prediction - are, according to Maturana, not necessary to constitute a genuine scientific practice.

2. *Objectivity and parentheses*

One of Maturana's often used premises is that in our common daily experience we can not distinguish between an illusion, a hallucination and a 'real' perception. This is what the frogs taught him. One can only make such a distinction by using another experience and treat it in one way or another as superior to the first experience, i.e. one has to use a criterion of authority. The distinction between real and false perception is a social activity. This leads Maturana "... to put a question mark on any perceptual certainty" (*Ibid.*, p. 331), and invite us to follow him on the Path of Objectivity in Parentheses.

The differences between the Path with and the Path without Parentheses are as follows:

Without Parentheses we assume that the *existence* of objects, relations between them, etc., is independent of the observer. The independent domain of existence which arises as a consequence of this attitude is usually called the '*universe*'. This independent domain furthermore determines '*truth*'. Objectivity-without-parentheses amounts to *reductionism*, because it is assumed that reality is one ultimate domain. Besides, it means that those who have access to this domain (in the past these were priests, nowadays scientists often take this social role) are necessarily right in every debate: "In the universum coexistence demands obedience to knowledge". (*Ibid.*, p. 332).

With Parentheses we accept that the existence of things is only 'brought forth' at the very moment at which an observer starts making distinctions. Thus there are as many *truths* as there are brought forth domains of existence. These domains arise from different sets of distinctions, and therefore they are independent of one another: there is no reduction (we will see that Latour speaks here of 'irreductionism'[3]). This also means that there is no universe, but that instead there are '*multiversa*'. And if there is a serious disagreement between observers, then it can only be solved by establishing a "... common versum through coexistence in mutual acceptance. In the multiversa coexistence demands consensus, that is, common knowledge." (*Ibid.*, p. 332).

3. Evaluation of Maturana's philosophy

"If there is no assertion and acceptance of the statement 'there is a problem', there is no problem."[4] Maturana's work in general shows that, for him, there are no longer any real problems in the theory and foundations of biology or psychology. His own story (Maturana, 1987) now and then nearly seems to burst out of the postulated parentheses, so large is his confidence in his own Objectivity-as-long-as-one-thinks-of-the-parentheses.

This leaves us, theoreticians of psychology, however, behind with some serious (at least) practical difficulties. First, it becomes pointless to read any more about theory and foundations of psychology, if we choose to accept Maturana's theory. We already saw how he discarded all other philosophy of science, and virtually all theories in biology and psychology undergo the same fate. Illustrative for the attitude that follows from this acceptance are two quotes from the author, in an interview, which is to my knowledge the only 'conversation' he ever engaged in. As the interviewer tells him that he hears a lot of Kuhn and Bateson in Maturana's words[5], Maturana replies: "So you don't hear me" (Maturana, 1985, p. 36). Besides this, Maturana remarks that he makes an uninterested impression upon other people.

In the second place, it is made pointless for us to *write* anymore, since to us as readers of Maturana there is only this one choice left: we accept or reject Maturana's theory. Here our problem becomes somewhat more fundamental, because we surely would like to enter a discussion with Maturana, for instance about the necessity of the primacy of biology over psychology, the role of 'emotions' and 'dispositions' in his psychology, his 'experiental' scheme of the scientific explanation, etc. This we cannot do for *in between* the parentheses which guard his articles and books, there reigns right enough the Objectivity and the Compulsion of the Argument.[6] Again the interview mentioned above provides an illustration. On a picture that goes with the interview, we see Maturana wearing a T-shirt that says: "Don't ask me any questions, I just might tell you the (truth)" (Maturana, 1985, p. 35).

These two consequences of Maturana's perspective we just have to reject, even if it were only because accepting them would reduce us to beggary. We therefore prefer the position of a therapist-according-to-Maturana. In the Mendez, Coddou & Maturana article, the notion of Objectivity-In-Parentheses is offered as the best attitude for the family-therapist. And since we, theoreticians of psychology, find ourselves in regard to psychologists and their theories as psychologists in regard to their subjects and clients, this attitude leads to accepting all perspectives as legitimate views of a particular situation. We must for the time being not reinforce one and reject others, but try to establish with our own presence a conversational praxis in which perspectives can be exchanged. Something like this should be our attitude in discussing the foundations of psychology. In other words: we are going to take Maturana's own words about mutual acceptance, multiversa, and consensus seriously.

In what follows, I would like to elaborate on this attitude by introducing an author from a seemingly remote nature: the French sociologist, philosopher, historian and anthropologist of science Bruno Latour.

4. Three turns in the study of science

Bruno Latour, as is perfectly clear from the quote on top of this paper, has not much patience with cybernetics. He became originally known for a study, undertaken with Steve Woolgar, of the 'real' life in a scientific working place: "Laboratory Life" (Latour & Woolgar, 1979). It turned out to be one of the most influential works in a series of studies in what became known as the '*empirical turn*' in the philosophy of science. In the book, subtitled "The construction of Scientific Facts", Latour & Woolgar argue that scientific facts arise through a social construction process, and not by the pure application of 'the scientific method'. This radical claim is further developed by Latour in a textbook for the British Open University (Latour, 1987). The starting point of this really breathtaking book of adventures is the pitiable position of the Outsider, who dares to dispute one of the celebrated dogmas of the new Established Church, modern 'technoscience'. This outsider position turns out to be untenable: the costs you have to make to attack an accepted fact are simply too high for an individual. This falsifies one of the most cherished beliefs of science: scientific statements are not verifiable by everyone. Partly from the money point of view the neutral social student of science chooses to study 'science in action', where the controversy about the claim that has to become a fact is still going on.

This is where the second, the '*sociological turn*' takes place. Science is no longer considered to be the only way of getting to a body of knowledge independent of culture and society, but - as so much else - a social activity through and through. Unlike other students of science, Latour does not try to develop a sociological theory of the development of science or knowledge; that would be a strange enterprise indeed: trying to be more scientific than scientists. No, when elaborating the strong claim that all knowledge is social[7], we can not invoke another 'scientific theory', even if this is now a sociological one. Instead of replacing one theory by the other, Latour wants to treat both, natural and social sciences, 'symmetrically'.[8]

This attitude amounts to the following, when it comes to the history of "The Pasteurization of France" (Latour, 1984/1988):

> "... instead of extending our knowledge of French nineteenth-century society to Pasteur's bacteriology with very disappointing results, and with a view merely to explaining the most superficial aspects of his science, it was easier and faster to suspend our knowledge of French nineteenth-century society and to follow, in the very technical aspects of Pasteur's benchwork, how a new social link was forged. Instead of imposing a far-fetched *implicit* social interpretation of their interests on the actors, this approach displayed the *explicit* translation by the Pasteurians of both a new society and a new science. The price to pay for such an approach was to give away the claim that sociologists and social historians know society well enough to explain the sciences. The price seemed to me a light one." (Latour, 1988b, p. 4)

The article from which this was extracted then tries to show that even a theory as abstract as Einstein's relativity theory is social through and through, including its technical details. For this purpose Latour launches a third, the *semiotic turn*. The bulk of the material of a social student or historian of science consists of *texts*: both theoretical as more empirical, but also 'co-texts' as books and lectures for the general public, newspaper articles, minutes, interviews, diaries, etc. By applying some 'semiotic tools' Latour thinks himself capable of showing, just by the way 'the story' is told, how political even Einstein is in his relativity theory.

And this enables Latour himself to elucidate in a new way the concept of 'context', of which so many sociologists of science state that it can explain the development of science.

5. *The problem of explanation as a way in*

As an illustration of what Latour's philosophy of science amounts to, and in order to show some striking similarities between Latour and Maturana[9], we take Latour's notion of explanation, i.e. 'the strong explanation' (Latour, 1988a). Such a strong explanation, with the 'power of deduction', Latour states, becomes necessary when one wants to 'act at a distance'. With the notion of a strong explanation, we immediately arrive at such notions as 'information' and 'control'. Information consists in those elements which, in a given situation x' in which the Explainer is *not* present, can be "mobilized, transferred, accumulated" (Latour, 1988a, p. 159) into the situation in which the Explainer *is* present (x). According to Latour, making strong explanations is the same as trying to build an empire, in which all the relevant actors (or 'actants') do exactly what you want them to do. (See also 6., point 2.) For Latour it becomes a *political choice* to give up strong explanations.

In this respect, Latour and Maturana could have an interesting conversation, in spite of Latour's prejudice against systems theory and Maturana's disinterestedness in other authors. Maturana also rejects the usual scientific practice of explaining, but in his case as a result of what he calls 'logical bookkeeping': according to Maturana, explanations take place in the domain of operational coherences in the praxis of living of the observer. And explanations do absolutely not take place in the domain of the phenomena, which are said to be the result of the explaining mechanism. The big difference between Maturana and other Explainers is that he refuses to build a network.[10] In this way Maturana builds his own home, and everybody can choose to use his building plans or not. Thus, the scientist Maturana escapes the criticism of Latour.

Conversely, Latour uses a kind of explanations that fit into Maturana's general structure of a scientific explanation (Maturana, 1987, p. 325 ff; see also 1., points 1-4.). In "The Pasteurization of France" for example, Latour wants to explain how Pasteur could become the immortal genius he is today (= 1. the phenomenon-to-be-explained). In order to do this, Latour introduces two mechanisms that made Pasteur immortal (= 2. the proposal of (a) generative mechanism(s)). In the first place there is not the genius of one man, who 'discovers' the microbes out of the blue, but instead, the translation and accumulation of forces and goals (interests) of very different groups - especially the big movement of hygienists - in such a way that everybody gets interested in Pasteur's laboratory: this is the primary mechanism. The second mechanism is also a matter of translation: all the success of this big alliance is reduced to the laboratory, the power of the scientist Pasteur.

Finally, Maturana's demand for extra phenomena (3.), which must reveal themselves by the same mechanism, is met with the explanation of the initial failing of the traditional medicine to fight epidemics: because of *lack* of allies. (4.)

6. *Infrareflexivity as a way out*

All sciences (also) dealing with human beings as their subjects have at least one problem in common: the problem of reflexivity. In this paper we encounter this problem in two forms. For Maturana the problem consists in the fact that biology deals with human beings and is at the same time a human enterprise. By putting his work in parentheses, Maturana was able to solve this problem. In 3. we argued that paradoxically this solution leaves others with no breathing space. Let us see what Latour has to say in this respect.

In Latour's line of work the problem is especially acute since the name alone, science of science, already exhibits the problem. According to Latour: "We want to be at once *more* scientific than the sciences - since we try to escape from their struggles - and much *less* scientific - since we do not wish to fight with their weapons" (Latour, 1988a, p. 165). This confronts us with three paradoxes:

1. How can we at the same time be 'here' and 'there', i.e. *within a text* and in that where the text refers to: this problem is common to all writing. All texts are *about* something other than the text itself. The problem is that every reference to elsewhere takes place only on paper, a fact that leaves the status of the reference quite insecure. Only if we resolve this paradox we can go on writing stories.

2. How can we at the same time be *here*, *there* (in the situation described) *and in between*, "managing the network that ties the two together" (*Ibid.*, p. 165). This is the problem common to all sciences and the solution of it would enable us to remain scientists. If we would like to maintain for instance that 'man is an information processing device', then we would have to be at the same time here in the text and there, at four billion other places (compare 1.); and besides this we would have to make those four billion others stay in their role and be decent information processing devices.

3. What middle course is possible between "being believed *too much* by the readers and *not enough*" (*Ibid.*, p. 166). This is the problem of the reflexive authors and we recognize it as one of the problems we encountered reading Maturana too: he is not looking for a middle course, but leaves us with the choice between rejecting and accepting his ideas. If we could solve this problem, we would be able to write "reflexive accounts" (*Ibid.*, p. 166), texts which take into account their own production, and in this way try to get their readers not to believe too much and not too little of what it says.

In other words: every text always deals *with something else*, even if this 'something else' is the text itself in the text. It is never *the same actor* who features in the text as the actor who wrote it (Latour). A reflection on the praxis of living doesn't change this praxis, but takes place in a meta-domain with regard to this praxis (Maturana).

The alternative way for the student of science is called by Latour 'infra-reflexivity'. Instead of trying to dazzle the reader with self-reference, Latour chooses self-exemplification, i.e. the text itself is an example of what it says. This means that Latour can not claim any privilege: "... my own text is in your hands and lives or dies through what you will do to it" (*Ibid.*, p. 171).

Still we have to solve, in one way or another, the problem that all texts are about 'absent elements', that is: they are written in A (the text) about B (that which A refers to). We choose not to take the scientistic way: make the connection between A and B, then forget about B and claim that B is now in A. In other words: the world is covered by the laws (mechanisms, theories) in the text. If you want to change this, you can take the side of phenomenologists or that of the ethno-methodologists: trying to act as if you stay in setting B. Or you can take the side of the deconstructionists or the meta-reflexivists: staying in A and not talking about anything anymore. Latour suggests the following alternative:

> "... why don't we take this activity as the name of our game: displaying the knower and the known and the work needed to interrupt or create connections between A and B ? I take this as the non-scientific way of studying both natural and social sciences; I also take it as a possible definition of infra-reflexivity." (*Ibid.*, p. 172)

It is this game between knower and known that we can also see as the theme of Maturana. Maturana constantly thematized the difference between the 'reality' of the living system on the one hand, and the domain of the descriptions and explanations of the observer on the other.[11] And although Maturana is still 'knitting' a network within and in between his own texts, he doesn't want to be imperialistic.

For the scientist of science Latour, the rejection of 'strong explanations' leads to a preference for 'one-off-explanations' and 'tailor-made causes'. And instead of aiming at a meta-language, he strives for equality with the studied discipline ('hybridization'). He also abandons the 'Machiavellistic strategy' for the study of science, by pleading to leave the academic circles:

> "... and to tie our work to the many struggles to resist being known, explained, studied, mobilized or represented." (Latour, 1988a, p. 175)

Our present concern was the foundation of science in general and psychology in particular and here too, such an approach seems promising enough to elaborate it further to make it fit this concern.

7. A plea for taking dialogue seriously

All in all Latour is leaving many of the time and again formulated pretensions of traditional philosophy of science, especially in the second part of his "Pasteurization of France" where he introduces his own ideology: 'irreductionism'. In the Spinozian form of a "Tractatus Scientifico-Politicus" he starts off with the proposition:

> "*1.1.1* Nothing is, by itself, either reducible or irreducible to anything else." (Latour, 1984/1988, p. 158)

After this he develops his own ontology (there are only potencies, forces, strengths/weaknesses, entelechies, monads or actants, which ever one prefers), 'sociologics' and 'anthropologics', in a style that is not only funny and beautiful, but very convincing too.

With his rejection of science as an extraordinary 'discourse', Latour takes a step which a number of others have made in recent years, partly under the name of 'relativism' in philosophy (of science). A considerable number of authors translated the negative claim in a more positive one by claiming that 'the dialogue' should be our major concern, and not 'truth', 'laws', or 'statistical significance', etc.

So for instance, the philosopher - with little p - Richard Rorty states that science is just one among many possible literary genres. And following Sellars he describes his 'pragmatic' philosophy as "... an attempt to see how things, in the broadest possible sense of the term, hang together, in the broadest possible sense of the term" (Rorty, 1982, p. xiv). And in an article on Rorty, the author summarizes Rorty's ultimate goal as follows:

> "Thus, the vision Rorty holds out for a 'post-philosophic culture' is one of creative inquiry freed from the bad faith evidenced in the search for ultimate foundations or final justification, a 'conversation' from which no one is excluded and in which no one, especially not the philosophers, holds a privileged position. The sole aim would be to keep the conversation going, or, as he puts it, 'to let a hundred flowers bloom'." (in Baynes et al., 1982)

The viewpoint examined here bears also resemblance to that of John Shotter. Shotter pleads for a 'non-cognitive, practical, social constructionist account' (e.g. Shotter, 1987) of social activities. Where most psychologists want to get at unequivocal definitions and theories, the social constructionist casts off this pretension, because in the human sciences we deal with words which stand for 'essentially contested concepts'. This expression is borrowed from Gallie, who defines them as concepts "the proper use of which involves endless disputes about their proper use on the part of their users" (Gallie, 1955-1956, p. 169, cited in Shotter, 1987).

This last point I would like to stress some more. In the human sciences - and I mean here: from biology to sociology, and already when we are defining 'humans' and 'sciences', there has never been any but local, temporal consensus about the use of even the most elementary concepts.[12] This being the case, it is important to realize that we can treat these disputes about definitions in two fundamentally different ways. We can, on the one hand, consider our own definition as the only right, rational, etc., and then regard all others who agree as the only rational partners in the ongoing debate. But we could also consider our definition as a 'work-away-definition' which can and need only serve its duties locally and temporally.[13] If we want to take the dialogical approach seriously, we have to take the second position.

Francisco Varela, co-author of a number of books and articles of Maturana is a proponent of a dialogical approach too:

> "The fundamental paradigm of our interaction with a control system is instruction and the unsatisfactory results are errors. (...) The fundamental paradigm of our interaction with an autonomous system is a conversation, and its unsatisfactory results breaches of understanding." (Varela, 1979, p. xii)

We see that Varela employs a very different style or tactic in introducing his ideas (which are basically the same as those of Maturana): he tries to tie his work to all thinkers who propose theories like his.

Other philosophers and scientists could be cited here as well.[14] The drawback of much of the resulting work of these authors is that it is stated in very general terms and that it sometimes falls into accusations of other authors who are considered trying to shut the dialogues down. The work of Latour on the other hand shows some examples of "affirmative action extended to the social sciences" (Latour, 1988a, p. 175), while constantly aware of the fact that the text always remains a form of 'talking about' (Kouwer, 1973). Maturana frequently warns us that we only have a language of objects at our disposal. And words like 'dialogue', and many more which featured in this paper, when written down, seem to refer to 'things'. However, this is not the same as an actual dialogue, in which the monologue of a text is at the most only the contribution of one partner.

All this taken together, we claim that it is necessary that we are seriously going to enter dialogues, to begin with our most far away antagonists. Such a method for the foundations and theory of psychology would give us the following advantages. Firstly, we establish the dialogue, which sometimes threatens to be shouted down by the 'monological character' of the text. Secondly, it is only within a dialogue that we can argue and show what such a different, dialogical approach would look like, that is, in terms shared by the other.[15] Thirdly, we could use our reader's position as a point of reference for the assessment of the propositions some psychological text consists in. Lastly, we could claim, following Latour, that besides fulfilling the requirement that we should have solved one or more of the problems of the partner, such a 'dialogue' would not have been completely successful if we didn't learn something from the conversation partner as well.

In a case study guided by the dialogical principles stated above[16], I entered in a conversation with the cognitive psychologist Ray Jackendoff (Meynen, 1989; Jackendoff, 1987). From this I learned that one can very unwillingly feel obliged to accept, due to scientific tradition and lack of known alternatives, some devastating things, namely that people are totally "biologically determined information-processing devices". On the other hand, I think I was also able to shed some light on problems of which Jackendoff himself declared that they were unsolved by his own perspective.

Notes

1. This paper is based upon an earlier unpublished manuscript (Meynen 1989). The author wishes to thank Arno L. Goudsmit for valuable comments.

2. The frog originally features in Lettvin, Maturana et al. 1959. For a summary, see Maturana 1983.

3. See Latour 1984/1988 (Second Part), and 7. of this paper.

4. This is the opening sentence of the Mendez, Coddou & Maturana contribution to the special issue of *The Irish Journal of Psychology* on Radical Constructivism, Autopoiesis and Psychotherapy. Mendez et al., 1988.

5. And these are certainly not the only names one could mention in this respect.

6. Compare the title of Maturana 1988.

7. The claim of the so-called 'Strong Programme', in which it indeed obtains the status of the only rational theory of knowledge. See for example the discussion between Bloor and Laudan: Bloor 1976, Laudan 1981 and Bloor 1981.

8. Besides Latour 1987, also Latour 1988a and 1988b.

9. More striking similarities between the two authors are to be found in their respective views on the status of (scientific) objects, and their origin.

10. For instance, he hardly ever cites someone else who has comparable views, but nearly always defines his own terms, most of the time with respect to each other.

11. See for example Maturana 1987, pp. 366-368.

12. It is one of the main issues of the study of science (starting with Kuhn, see for example also Collins 1985), to show that the same applies to the purportedly 'harder', natural sciences.

13. Qualification first used by J. Quackelbeen, 1990, *Wegwerktekst*, Gezelschap voor Psychoanalyse en Psychotherapie, Ghent.
Compare also the two Paths of Maturana, 2.

14. Let me name another two: the Dutch psychologist B.J. Kouwer (1927-1968), whose major works unfortunately are not available in English (see Kouwer 1973 en 1963), and - more recently - Kenneth Gergen.

15. So maybe we even have to discard the term 'dialogue' from time to time, and replace it by communication, interaction, critique, discourse, controversy, or still something else.

16. Hopefully the reader has already noticed that this paper is a 'case study', or an attempt at 'self-exemplifying' too.

References

Baynes, K.; Bohman, J.; McCarthy, T., 1987, *After Philosophy. End or Transformation*, MIT Press, Cambridge (Mass.).

Bloor, D., 1976, The Strong Programme in the Sociology of Knowledge, *Knowledge and Social Imagery*, D. Bloor, Routledge, London, pp. 1-19.

Bloor, D., 1981, The Strength of the Strong Programme, *Philosophy of the Social Sciences*, vol. 11, pp. 199-213.

Collins, H.M., 1985, *Changing Order. Replication and Induction in Scientific Practice*, Sage, London.

Gallie, W.B., 1955-56, Essentially Contested Concepts, *Proc. of the Aristotelian Soc.*, vol. 56, pp. 167-198.

Jackendoff, R., 1987, *Consciousness and the Computational Mind*, MIT Press, Cambridge, Massachusetts.

Kouwer, B.J., 1963, *Het spel van de persoonlijkheid*, Bijleveld, Utrecht.

Kouwer, B.J., 1973, *Existentiële psychologie. Grondslagen van het psychologisch gesprek* (posthumous), Boom, Meppel.

Latour, B., 1984/1988, *The Pasteurization of France*, Harvard UP, Cambridge, Massachusetts/London.

Latour, B., 1987, *Science in Action. How to Follow Scientists and Engineers through Society*, Open University Press, Milton Keynes.

Latour, B., 1988a, The Politics of Explanation, in *Knowledge and Reflexivity*, S. Woolgar, Sage, London, pp. 155-176.

Latour, B., 1988b, A Relativistic Account of Einstein's Relativity, *Social Studies of Science*, vol. 18, no. 1, pp. 3-44.

Latour, B.; S. Woolgar, 1979/1986, *Laboratory Life. The Construction of Scientific Facts*, Sage UP, Los Angeles/Princeton.

Laudan, L., 1981, The Pseudo-science of Science ?, *Philosophy of the Social Sciences*, vol. 11, pp. 173-198.

Lettvin, J.Y.; Maturana, H.R.; McCulloch, W.S.; Pitts, W.H., 1959, What the frog's eye tells the frog's brain, *Proceedings of the IRE*, vol. 47, pp. 1940-1951.

Maturana, H.R., 1983, What is it to see ?, *Arch. Biol. Med. Exp.*, vol. 16, pp. 255-269.

Maturana, H.R., 1985, An interview with Humberto Maturana, *The Networker*, May-June 1985.

Maturana, H.R., 1987, The biological foundations of self consciousness and the physical domain of existence, in *Physics of cognitive processes*, E. Caianiello (ed.), World Scientific, Singapore, pp. 324-379.

Maturana, H.R., 1988, Reality: The Search for Objectivity or the Quest for a Compelling Argument, *Irish J. of Psych.*, vol. 9, no. 1, pp. 25-82.

Maturana, H.R.; Varela, F.J., 1980, *Autopoiesis and Cognition. The Realization of the Living*, Reidel, Berlin.

Mendez, C.L.; Coddou, F.; Maturana, H., 1988, The Bringing Forth of Pathology, *The Irish J. of Psych.*, vol. 9, no. 1, pp. 144-172.

Meynen, H.T., 1989, *De Avonturen van een Lezer - over reflexiviteit in de studie naar de grondslagen van de psychologie*, Rijksuniversiteit Groningen, Faculteit PPSW, Vakgroep Psychologie, unpublished manuscript.

Rorty, R., 1982, *Consequences of Pragmatism*, Essays 1972-1980, Minneapolis Unuversity Press, Minneapolis.

Shotter, J., 1987, Vico: the Role of Rethoric and the Concept of the Imaginary in the Making of Social Life, Paper presented at the Meeting *Problems of Meaning in Modern Psychology*, Manchester Polytechnic, 14-3-1987.

Varela, F., 1979, *Principles of biological autonomy*, North Holland Elsevier Company, New York.

Von Foerster, H., 1981, On Cybernetics of Cybernetics and Social Theory, in *Self-organizing Systems: an interdisciplinary approach*, G. Roth and H. Schwegler (eds.), Campus, New York/Frankfurt.

A ONE-SIDED BOUNDARY: ON THE LIMITS OF KNOWING ORGANIZATIONAL CLOSURE

Arno Goudsmit

University of Groningen
Department of Psychology

Abstract

Maturana's theory of autopoiesis, and particularly his ideas on language and the ensuing construction of reality, are used as the major line of thought in this contribution. Some basic notions from his theory are used and extended as a framework to provide some explicit thoughts about what is considered as a particular omission in his theory: an affirmation of the impossibility to empirically observe the internal states of an autopoietic system. It is maintained that the definition of autopoiesis entails a concept of 'state' that coincides with 'state transition'. A researcher who is interested in observing the internal states of an autopoietic system will become enmeshed in the impossibility to perceive his own perceptual acts. This is related to Merleau-Ponty's notion of a 'negative philosophy', i.e. a philosophy in which the central theme is the impossibility for consciousness to grasp itself as a consciousness of the world. Finally, the concept of a 'one-sided boundary' is presented as a metaphor for the inaccessibility of closure. It is maintained that entering a closed system is not so much a matter of observation, but rather of its cessation.

1. Introduction

1.1.

Maturana has elaborated the theory of autopoietic systems with respect to language, perception and the construction of reality (e.g. 1987). One attractive feature of this theory is that it is concerned with the autonomy both of the observed system and of the observer. In this contribution it will be explored whether an observer, as an autopoietic system, is able to empirically observe the autopoiesis (and more generally: the organizational closure) of another autonomous system. We will find that the answer is: no. However, this does not mean that we are to abandon the theory of autopoietic systems, but rather that it is to be

G. Van de Vijver (ed.), New Perspectives on Cybernetics, 175–205.

complemented with a delineation of its own limits, and that it is by knowing these limits that we are able to find our way to the 'closure' of autonomous systems.

We will follow Maturana's theory, and extend it at times with additional concepts. This will be done in sections 2 through 6. Next, in sections 7 and 8 we will use the presented ideas in order to describe the impossibility for an observer to perceive the organizational closure of an autopoietic system, and see what fruits these limitations bear.

1.2.

Autopoietic systems exist as networks of processes, that continuously coincide with the processes by which they are generated. The processes that take place in the autopoietic system are the processes by which the system is kept alive, i.e. the processes by which it continues to produce itself.

The autopoietic system is defined such as to produce itself as a self-producing and self-preserving network of relations between processes and components. This definition does justice to a fundamental property of living systems, namely that they are capable of conserving their capacity to survive, *by* surviving. Living systems succeed in this by their capacity to adapt to environmental changes. This adaptive capacity is both the way in which they survive, and that which they preserve in surviving.

Therefore, the production process in autopoietic systems is not defined as separated in a strict sense from that what it yields, the 'product'. We encounter in the definition of autopoiesis a peculiar merge of, on the one hand, a network of relations between processes and components that together make up the 'structure' of the autopoietic system, and, on the other hand, the processes that yield this network. This merge of a product and the process that produces it, can be described as a merge between a system's states and its state transitions (cf. 1.3).

1.3.

The autopoietic system understood as a self-producing network of relations between processes and components, does not fit in with the regular definitions of systems (e.g. Minsky, 1967). A *state* of a regularly defined system is a particular set of dispositions of the system to respond to new inputs. In empirical investigations, a system's state is usually considered as a particular cluster of dimensions that can be assigned to the structure of the system under observation (one dimension being for example an object's weight). Dependent on the particular state and the particular input that the system receives, a particular *state transition* may take place. As a result, the system obtains a new state according to which it may respond to the same inputs in an entirely different way. These notions of state and state transitions, as used in the context of a regular system, are of different types: a state transition being a predicate of two states.

However, for autopoietic systems this logical typing does not hold. A state of an autopoietic system is at the same time a particular state transition. This merge of state and state transition can be described as the merge between an operand and an operator that operates upon it. This treatment of an operator as its own operand has been discussed e.g. by Kauffman & Varela (1980, p. 195).

The autopoietic system exists as long as it succeeds in using the effects of its ongoing interactions as parts of its own dynamics of state transitions. Precisely at the locus of its interactions with the medium (cf. 2.4) in which the system exists, its dynamics of states is decisive for its adaptation. Its state at a particular moment *is* the way in which it manages to deal with perturbations. Or, formulated more elliptically: the state of the system at a particular moment is its particular state transition at that moment that takes place as a result of the interactions the system is having at that moment. It is in its adaptive capacities that the system's survival mechanism resides, that is: its adaptive capacities consist of the preservation of the merge of states and state transitions. If we are to know empirically how the adaptive capacities of an autopoietic system work, it must be by knowing its states as state transitions during their interactions.

1.4.

Because, with respect to autopoietic systems, the categories of 'state' and 'state transition' are not of a different logical type, any stability that is observed in an autopoietic system can only artificially be named a 'state'. The artificiality is situated in the fact that the observer 'opens' the system's closed network of operations. That is, he describes a stability in the system's structure and he ignores some vital interrelationships between processes in the system that take place at the same time. Were he to take them into account, no state could be described at all, and no explanation could be given of the system's behaviour.[1]

1.5.

It is this impossibility for an observer to have access to a system's domain of states (cf. Maturana, 1982, p. 289) that justifies the idea of a 'logical bookkeeping' (cf. Varela, 1979, p. 12) between a 'domain of descriptions' in which the observer expresses that which he sees, and a 'domain of operations' in which the system's dynamics of states takes place. Each attempt to describe the domain of operations of an autopoietic system necessarily fails. This point, in my view, is too little emphasized by Maturana, Varela and others. Though Maturana does make a general remark about the impossibility to know a system's state (1987, p. 343, p. 350), and though Varela (1983, p. 157) maintains that structural changes are difficult to observe at the level of neurons, the major principal point is not made, which is *that the very idea of autopoiesis (or rather organizational closure) entails an impossibility to describe its very closure in exact observational terms.* Whenever we want to describe something, we need substrates and predicates to be definable and observable in *some* way. In the case of autopoiesis, no proper predicates (state transitions) and no proper states (substrates) can be assigned to the system without ignoring its merge between states and state transitions, i.e. without opening the system's closed network of operations, and 'explain' it into a stratified order.

The observer, on the other hand, is bound to his use of a domain of descriptions in which states and state transitions are used as distinct logical types (substrates and predicates), even though these types are of no help to him in describing the autopoietic organization. It is only by using them that descriptions are possible at all and that objects can be observed at all.

The domain of operations of an organizationally closed system lies beyond the scope of what can be described and observed. The closure of such a system's operations means that these operations, the coinciding of states and state transitions, do not exist as observable phenomena. This inaccessibility to observation appears to us as a barrier beyond which empirical methods seem to fail. It is as if we can think only in highly abstract terms about that which is beyond that barrier, without being able to perceive it. We are reminded of Kant's division between noumenal and phenomenal worlds. Is the organizational closure of such a transcendental status ? We are not to maintain that.

1.6.

Instead, we are to conclude (cf. 8.2) that the closure of an autonomous system exists in a domain that is inaccessible to us in the same way in which our own pre-reflexive experiences are inaccessible to conscious inspection. It exists in a domain that is the basis of our own bodily existence, and which we, following Madison's (1981) reading of Merleau-Ponty (1964) may consider as 'transcendental in immanence'. It will be maintained that language (and being an observer) is an impediment to our access to organizational closure (cf. 8.5).

2. *First order distinctions*

2.1.

The first order distinctions the organism makes during its interactions with its medium exist as primary qualities of its praxis of living (pleasant/unpleasant, hungry/satisfied, etc.). They are the activities by means of which the autopoietic system compensates the perturbations it undergoes, e.g. by the activity of digestion a particular distinction is made between 'hungry' and 'not hungry'. This kind of distinction exists in this activity and comes into existence through it.

By making such distinctions an individual does not yet distinguish between:

'the map' versus 'the territory'
'me' versus 'you'
'form' versus 'content'
'subject' versus 'object'
'process of distinction' versus 'outcome of distinction'[2]
'impression' versus 'expression'
'inside' versus 'outside'
(see also 3.4)

2.2.

When an individual is making a distinction, this is not to say that he discerns some figure against some background. Rather, it means that there is something that an external observer could see as an object in the observed individual's environment, which the latter is distinguishing *without* seeing it as an object. He is *making* the distinction by his behaviour, but he is not *having* the result of the distinction as a kind of object. Examples of this are easily found in the behaviour of animals and young pre-linguistic children. The ease with which persons or things are recognized does not allow the external observer to conclude that the observed individual is able to distinguish the recognized person/thing as an object in the real world (the object as conceived by the external observer), but only that the observed individual has acquired a particular capacity to deal with his environment (as the external observer sees this environment).

2.3.

Let us use the term '*quasi-object*' as different from 'object': 'object' refers to that what is observed in language by second order distinctions (see 5); 'quasi-object' refers to that what is distinguished by a first order distinction. A quasi-object, therefore, does not exist for the individual (to be called the *quasi-observer*) as an object; it can only be described by an external observer as the object that the latter ascribes to a non-observing individual as if the latter were also observing it and having it as an object. This is the case for example when we observe a cat chasing a mouse. Then the cat is ascribed the perception of this mouse, as if the cat had made a second order distinction (cf. 5), and had observed the mouse as an object. Rather, in fact, the cat can be observed to act in such a way that by his behaviour he distinguishes the mouse. The mouse can be said to be a quasi-object for the cat. It is *as if* the cat sees the mouse. The term 'quasi-object' is useful for a proper understanding of language (see 5.3).

2.4.

The same difference between the positions of the observer and the observed holds in respect of the terms 'environment' and 'medium'. The environment in which the observed individual is seen to make his distinctions, is the environment as it exists for the external observer. For the milieu as it exists for the individual himself, Maturana (1987, p. 359) uses the term 'medium'.[3]

3. *Experiences as first order distinctions*

3.1.

Experiences exist as properties of the interactions the system has with its medium. Experiences exist as pre-reflexive modes of 'being in the world'. No objects are observed or constructed. In this text, a strict distinction is made between the notion 'experience' on the one hand, and on the other hand the notions 'observe' and 'perceive'. Only the latter involve the (coming into) existence of objects.

Experiences do not specify objects in the system's environment, though an external observer might describe objects with what he sees the system interact. Then these objects have the status of quasi-object for the system that is having the experience. Experiences are not 'internal' or 'mental' representations of a quasi-object. However, an external observer may be convinced that the experiences of the observed system are the internal or mental correlates 'of' a particular object in the 'real' world. The observed system is then said 'to experience this particular object'.

However, experiences do not exist as internal representations of external objects. Instead, they exist in a domain where the experiencing individual does not distinguish between his own inside and outside. Therefore, the 'inside' of this individual in which the experiences are presumed to dwell according to the external observer (i.e. a researcher or a therapist), does not exist as such for the observed individual himself. The same holds for the 'outside' of this individual in which the objects are presumed to dwell according to the external observer. So both an internal representation and an external objective reality exist only in the observer's domain of descriptions, not in the domain in which the observed individual's experiences exist.

3.2.

Merleau-Ponty (1964) calls the Flesh (la Chair) the ontological substratum out of which the subject-object distinction evolves, but which is itself beyond this distinction. The Flesh itself remains beyond our scope of objectivity. It cannot be objectified.

We may consider that what Merleau-Ponty calls 'Flesh', as the domain in which experiences exist. Experiences are our mode of functioning, by which we participate in the Flesh. It is in our experiences that we exist before and beyond the subject-object dichotomy and that we are able to have direct, i.e. immediate contact with others.

3.3.

The first order distinctions we discussed above (see 2) as modes of functioning, by which the individual does not (yet) come to perceive objects, but nevertheless seems able to cope adequately with his quasi-objects, can be interpreted as the *experiences* this individual has.

In the domain of experiences, there is not so much an objective-subjective dichotomy, but rather an absence of this dichotomy. Only when language arises does this dichotomy come into existence. We will return to this difference between the domains of experiences and of language in 5 when discussing the origination of language, and in 8.4 when discussing the boundary of an autopoietic system in terms of a language barrier.

3.4.

Experiences and cognitions are indistinguishable from each other. This is a different notion than Ciompi's (1982) idea that feelings and cognitions are always going together, inseparably. Feelings and cognitions, according to Ciompi are not identical. According to the position I am taking here, experiences (feelings) and cognitions are identical at this level of functioning. Experiences are an elementary mode of cognition. For that reason, we may append the list given in 2.1 with a distinction that is not yet made either in first order distinctions: 'that which is experienced' versus 'the way in which it is experienced'.

At this place we cannot do justice to the wide range of possible experiences individuals can have. According to our use of terms, both the bending towards the light by plants, as well as the sexual attraction between persons is characterized in this text in terms of 'first order distinctions'.

4. Structural coupling

4.1.

When two systems, living or non-living, enter into an interaction, then this interaction leads to a continuous change of states in each system. Each system perturbs the states of the other system, and each compensates for these perturbations by means of a change of its states. Maturana uses also the terms 'co-ontogenic structural drift' (1988, p. 47) and 'ontogenic structural drift' (1987, p. 344). Thus two interacting systems yield changes of states that are mutually influencing. Due to this mutuality these changes of states lead to a structural isomorphism in the two interacting systems (Maturana, 1982, p. 290). Two round discs of glass, rubbed along each other, will yield the deformation of each, one becoming convex, the other concave. Such a development is spontaneous, and it takes place until the nature of the interaction changes in a way that at least one of the interacting systems disintegrates (e.g. the concave slice becomes too thin and breaks). For the sake of completeness, an established coupling can be dissolved, e.g. when the two interacting systems drift apart.

4.2.

When two living systems structurally couple, then their interactions lead to structural perturbations in each, that can be compensated by further behaviours of the interactants. This active capacity for adaptation is essential to life. Like non-living systems, the interactants develop a structural isomorphism at their sites of interaction, so that a particular structural fit arises between them. It is at this interface that the living system differs from the non-living: by the incorporation of perturbations in the living system's current dynamics of state changes.

If the perturbations triggered by the interactions between the interactants cannot be compensated by one of them, then that damaged system looses its capacity to survive and disintegrates. Then there is no longer a structural coupling between the two systems. Violence is also to be understood in these terms.

4.3.

During structural coupling between living systems, the interactants behave in a way that becomes coordinated, due to the development of their structural isomorphism. The coordination of these actions is a spontaneous outcome, not one that is intentionally sought for. It is a mode of growing together.

These coordinations of actions constitute distinctions that are consensual to the two interactants. These consensual distinctions are shared experiences. Since they are generated by a structural coupling between the two participants, the shared experiences take place as a shared activity. No objects are distinguished by consensual distinctions, though an external observer might find a correspondence between the interactive patterns and some object as he observes it in the environment of the two interacting systems. Then this object exists as a quasi-object for the interactants, as in 2.3.

5. Language and naive objects

5.1.

An observer who watches two individuals (say, two animals) behave in a domain of consensual distinctions, is often inclined to find the object to which this consensual behaviour seems to refer. Then this object, which we called a quasi-object for the interactants, comes into existence for the observer because he takes the behaviours of the observed individuals as tokens. By following these 'tokens', the observer comes to observe the object to which they seem to point. It is this capacity to take consensual behaviours as tokens for some object, that is according to Maturana the vital capacity to create a reality. Observing is equivalent to constructing ('bringing forth') the observed object.

5.2.

This capacity is a particular human affair. We are human by our ways of treating our interactive behaviours as symbols for something else. The meaning of these symbols is again a matter of consensus. So objects arise through a consensual use of symbols, which in their turn are consensual actions of the participants.[4]

The reason for denying the so called 'animal languages' and also the human 'body languages' the status of a genuine language, is that it is only in the eyes of an external observer that the performed consensual behaviours appear as tokens or symbols for some object. For those involved in the 'non-verbal language', however, this symbolic status is not there. Only some higher mammals have been observed to use particular sounds or behaviours to denote a particular danger, such as a dangerous animal, and only for those restricted situations their behaviours seem to have a consensually assigned meaning.

5.3.

In a language, the consensual pattern of actions itself *becomes a quasi-object* for the interacting individuals in regard of which they can coordinate their actions. That is, for an external observer it may seem that the interacting individuals coordinate their actions in respect of their own previous interaction patterns. For this reason this is called '*second order coordination*'. Here a recursion in the consensual coordination of actions arises, or, equivalently, a recursion in the consensual distinctions made by the interactants. To an external observer it then seems as if they are consensually distinguishing their own previously made consensual distinctions. Hence the term '*second order distinction*' is also used.

The notion of a quasi-object is particularly useful for us in order to be able to describe an individual's acts of distinction. Then 'the thing that is distinguished' can be indicated by the external observer as a particular quasi-object. This allows us to speak of second order distinctions as distinctions of (first order) distinctions, without the latter being observable for the observed individual himself.

5.4.

These second order consensual distinctions are useful as soon as the individuals no longer coordinate their simple actions (i.e. share their experiences) in a self-evident way. Consensually managing the occurring divergences of experiences is a second order coordination of actions. The behaviour by which this takes place is *language*.

Language is a (second order) coordination of the (first order) coordinated actions. As long as experiences are shared unproblematically, no language and no objects arise. Likewise, when in the course of a second order coordination of actions the first order coordination of actions no longer needs to be coordinated (that is: when the interactants (again) start to share their experiences unproblematically), then language disappears (cf. 6.7.ii)). Conversely, when the (first order) coordination of actions (i.e. the shared experiences) becomes a quasi-object for the interactants, it is because this coordination of actions is no longer taking place smoothly, and is distinguished. Their discoordination becomes noticeable and through it the object becomes perceivable, like the difference between the sounds of two strings, e.g. of 440 Hz and 441 Hz becomes noticeable as an interference pattern.

5.5.

Maturana gives the example of the dog and the man, playing a game in the park. The man throws a stick, the dog chases it and brings it back. We may understand this as a basic pattern of interactions, a first order coordination of actions that goes together with a shared experience (a consensual distinction). Then an interruption happens. The man starts a conversation with an acquaintance. The previous coordination of actions with the dog is no longer warranted. Then the dog barks. Now this barking, according to Maturana, is a case of language.[5] The barking is a way to deal with the disrupted first order coordination of actions (the game). If the man is to respond to the barking, then a second order coordination may arise between them that leads to a consensual distinction of the disrupted game. Since the game was a coordination of actions, this new 'game' is a second order coordination, i.e. a coordination of the previous game. Then the barking, as well as the man's response to it, become ways to make the previous game a signifier that refers to the stick that is to be thrown. Notice that the discoordination of actions (the disrupted game) is a circumstance due to which a second order coordination becomes possible at all.

This discoordination is not mentioned in an in other respects strikingly similar line of argument we encounter in Shotter's presentation of some of the basic ideas of Vico, concerned with the origination of meaning. The situation discussed is one in which primitive people are imagined to be confronted with a thunderstorm: "... as everyone runs to shelter from the thunder, all in a state of fear, an opportunity exists for them to realize that it is the *same thing* that they all fear; and a look or a gesture will communicate this. What we might call a 'moment of common reference' exists between them. (...) And here it is the fear shared in common which provides the first fixed reference point which people can 'find again' within themselves and know that others 'feel the same way'." (Shotter, in press, pp. 8-9).

Thus there arises "... a sensuous totality linking thunder, with the shared fears to do with the limits of one's being, and with recognizing the existence of similar feelings in others because of shared bodily activities. It is created, not out of a heterogeneous amalgam of events, but *within* a developed and developing totality of relations between people. (...) Thus, these first anchor points are to do, not with 'seeing' in common, but with 'feeling' in common, with the 'giving' or 'lending' of a shared *significance* to shared *feelings* in an *already shared* circumstance." (Shotter, in press, pp. 9/11).

5.6.

In both examples, there is an ongoing activity between two or more participants that becomes indicative of something beyond the current activity.

Notice that, by taking language as a process of (second order) consensual distinctions of (first order) consensual distinctions, Maturana does not so much account for the origination of an established set of signs and symbols with a relatively fixed denotation in a particular culture. He does not describe the role of cultural conventions in the signification of signs. Rather, his theory is concerned with the first origination of shared meaning. Likewise, it was Vico's conviction that the language that is arising as a shared significance to shared feelings in a shared circumstance, is primarily of a *poetic* kind.

5.7.

The objects thus generated are what we will call *naive objects*. The term 'naive' denotes the absence of a distinction between the object's form and its content. Thus, a person who perceives a naive object does not distinguish its form from its content. Alternatively we may call such a person a *naive observer*.

6. Naive and critical objects: perception and conception

6.1.

When a second order coordination (which is a second order consensual distinction) occurs, then that which was already performed by the interactants (the first order coordination of actions, which was a first order consensual distinction) is now made into a 'token' for that which is newly emerging.

When the experiences (which are first order consensual distinctions) become tokens for the newly emerging (naive) objects, this is to say that these experiences themselves now become consensually distinguished (by consensual distinctions of a *second order*). They become the (abstract) *form* of the naive object that comes into existence as the *content* of a conversation (however simple or poetic this conversation may be). This naive object's *form* (or its meaning, underlying pattern, essence) is not perceivable for the naive observers themselves. It is a quasi-object for them, i.e. only a meta-observer may perceive it as an object that is 'immanent' in the naive observers' perceptions. These naive observers may be said to perceive this form only as quasi-observers, in the sense in which the non-linguistic quasi-observers were said to perceive the naive objects they were distinguishing (cf. 3.1).

In the perception of the naive object the form of this object does not exist for the observer as a distinguishable entity as such. No formal aspects can be described by the observer, i.e. no aspects that abstract from the particular object under observation. In other words: the naive object does not possess synthetic properties. Its properties are analytic. They are perceived as 'belonging' to the object, and not as 'assigned' to it. The form-content distinction itself does not yet apply to the naive object. It does not have a form (as distinct from the content) for the naive observer.

At the level of the naive object no form can be observed by the (naive) observer. The form of the naive object does not exist for him, it is only a quasi-object. Both his experience and the naive object's form do not exist as such for the naive observer. When he does come to see the naive object's form (see 6.6, 6.8), it is by abstracting from his first order distinctions that constituted the experiences through which he came to see the naive object.

6.2.

When an observer is able to conceive the form of an object, i.e. when he is able to describe the object's form in terms that in some way abstract from its contents, then let us call this observer *critical* and let us call the object the form of which he thus conceives the *critical object*. The difference between the naive and the critical observer is that the former only perceives an object and the latter also conceives the object's form (or, what amounts to the same: its essence, underlying pattern, or meaning).

When an object is conceived as a critical object, then it consists of a *substrate* to which *predicates* are assigned. That which remains perceivable for the critical observer as naive objects are its *contents*. These contents then are perceived as a symbol or image of the *form* that is conceived. The form is considered either as the substrate or as the predicate (cf. 6.4).[6]

6.3.

Critical objects come into existence in a way analogous to naive objects. Like the naive object is made as a modus between individuals to deal with somewhat incongruent first order distinctions (experiences), the critical object is established as a purportedly objective point of agreement, e.g. between two conflicting parties (or within one person who perceives two more or less conflicting naive objects).

Conflict situations have at least two parties that have more or less incongruent practices (or one person having two more or less incongruent practices). The way in which naive objects are perceived may vary with the practices in which they are brought forth. In conflict situations that concern a particular naive object, the two parties may come to agree that the issue of their disagreement is a matter of diverging 'appearances' (the predicates) assigned to an 'essence' (the substrate), about which no disagreement exists. Or the two parties may come to agree that the issue of their disagreement is a matter of diverging 'meanings' (the predicates) assigned to 'symbols' (the substrate), about which no disagreement exists. Then the conflict is settled by splitting the topic of disagreement into a 'constant' part about which no disagreement exists, and a 'variable' part about which the conflict may continue.

But now the conflict is no longer about 'what is it' (substrate), but rather on 'how is it' (predicates). The agreed part then is the substrate (at times called 'essence', 'real thing', 'objective world', 'thing in itself'); the disagreed part is the predicate (at times called 'phenomena', 'subjective world', 'thing for us', 'meaning'). The substrate then becomes the new issue of agreement, the predicates may remain topics of disagreement.

6.4.
This can happen in one of two ways:

A: agreed: conceived substrate (essence, *form*)
disagreed: perceived predicates (appearances, *contents*)
B: agreed: perceived substrate (symbols, *contents*)
disagreed: conceived predicates (meanings, *forms*)

ad A.
Example: Two observers agree that a particular animal appears to one as a horse, to the other as a pony. Now 'horse' and 'pony' are (disagreed) perceived predicates to the (agreed) conceived substrate 'animal'. This conceived substrate itself is not perceived.

ad B.
Example: Two observers disagree about the interpretation of a particular laboratory animal's behaviour. Now both assigned meanings are (disagreed) conceived predicates to the (agreed) perceived substrate 'behaviour X'. The conceived predicates themselves are not perceived. An example of this kind can be found in Milner (1971, p. 168): "Cats also gave "pain suggestive" responses to stimulation in the superior and inferior colliculi". In this fragment the laboratory animal's pain is not at all evident for the investigator, who only seems to be able to observe behaviours that (as a kind of symbol) can be interpreted as denoting pain!

In case A the substrate is conceived, not perceived, whereas the predicates are perceived as naive objects. Conversely, in case B the substrate is perceived, the predicates are conceived. Each participant then is allowed a particular predication of the substrate. (Though the interactants do not need to go so far as to permit each other a right of possessing an individual 'internal representation', this type of substrate-predicate split prepares the ground for it (cf. 7.2)).

The two cases A and B differ in one vital respect:
in case A the critical object's form is its substrate (essence)
in case B the critical object's form is its predicate (meaning)

	form:	contents:
case A:	substrate= essence	predicate= appearance
case B:	predicate= meaning	substrate= symbol

In each case, that what is observed naively is called the contents; that what is not perceived naively is called the form. No other cases are possible, or it be the case in which cases A and B coincide: a thing's essence being equal to its meaning, of which its appearances are symbols. The description of such a 'thing', however, is either of type A or of type B.

6.5.

A naive object generally arises as a way of coordination between individuals concerning the first order coordination of actions they are having. A critical object generally arises as a way of coordination between individuals concerning the second order coordination of actions they are having, that is, concerning their naive objects.

Naive objects result out of situations in which experiences are not shared unproblematically (cf. 5.4); critical objects result out of situations in which naive objects are not shared unproblematically (cf. 6.3). In those situations, the interacting persons disagree about a particular naive object they are discussing.

Both in the case of naive objects and of critical objects, the resulting topic of agreement functions as a shared topic of orientation. Naive objects come into existence as the coordination of experiences; critical objects arise as a coordination of naive objects. One is tempted to think of the need for an 'Archimedean point', as in cartesian philosophy, as a generalized way of establishing a shared topic of orientation, a point of reference for persons when coordinating their lives, as e.g. in conflict resolution.

6.6.

Like the naive object comes into existence by taking the first order distinctions as a token, i.e. by (second order) distinguishing this (first order) distinction, so also the critical object's form may come into existence as a naive object by taking the second order distinctions as a token, i.e. by (second order) distinguishing them as plain first order distinctions. Then the form, in its turn, becomes perceivable as a naive object.[7] When this happens, it is not perceived as something that in its turn has a form and a content. It is what is usually called a 'gestalt', for its components are not perceived individually, but as a comprehensive entity.[8] As a critical object, however, it was not perceivable as a gestalt, it was only conceived as a pair of substrate and predicate.

When the form becomes perceivable as a new naive object, it becomes *known from*[9] the original naive objects by *incorporating* them, i.e. by starting to use them as tools (instead of seeing them as objects) for seeing.

The original naive objects then become the contents of the new critical object; but these contents are no longer perceived as the independent naive objects they originally were. They are no longer perceived at all in ideal circumstances (cf. 6.8). They are abstracted from. To the extent that they still remain perceivable, it is at the cost of the critical object's form being perceivable as a naive object.

The emergence of the naive object's form as a new naive object is usually described in terms of gestalt perception. As the new gestalt becomes perceivable as a naive object, it is due to the fact that the naive observer's original first order distinctions are no longer used as tokens. The gestalt is perceivable by integrating its constituents.

6.7.

Let us recapitulate these levels of distinctions:

i) A pre-linguistic individual (quasi-observer) uses only first order distinctions. These are his *experiences*. An external observer might find some relation between these experiences and something (the quasi-object) in the environment in which the individual is observed to exist.

In terms of coordinations of actions: experiences are shared with other individuals. Remaining conflicts consist of not (first order) coordinating actions (physical violence). These remaining problems are dealt with by a second order consensus.

ii) A *naive observer* uses second order distinctions in order to distinguish his first order distinctions (experiences), i.e. in order to use his first order distinctions as tokens for the naive object. That is, he uses his experiences as tools by means of which the naive object can be observed. The naive object is observed as *having a particular character*. Its predicates are only analytic properties.[10]

In terms of coordinations of actions: experiences are not entirely shared with others. Naive objects are a solution for this, as a way to manage the divergence between two persons' non-identical experiences. Remaining conflicts concern disagreements about the naive objects. These remaining problems can be dealt with by bringing forth critical objects.

To the extent experiences (first order distinctions) are (again) shared unproblematically between persons, the use of naive objects disappears. Then no incongruence between the behaviours of the interactants is distinguished any more, and their first order coordination of actions goes (again) smoothly.

iii) The critical object is conceived as *consisting of a pair of substrate + predicates*. The critical observer differs from the naive observer in that he not only observes the (old) naive objects, but also arranges them into a pattern that he can conceive, but not really perceive, between them. This pattern is either a new substrate (the 'essence') or a new predicate (the 'meaning').[11]

The critical observer therefore distinguishes both his first order and his second order distinctions. The former remain, as they were, tokens for naive objects (experiences through which naive objects could be observed); the latter become tokens for the pattern between these naive objects, though this pattern does not appear as a naive object (as in 6.6, 6.8), but rather in an imaginary way (cf. 7.2). The second order distinctions cannot (yet) be used as a tool (as were the first order distinctions) for seeing a naive object, but rather they appear as a 'text' of symbols that has to be decoded (cf. Shotter, 1984, p. 212). These symbols, as naive objects, are considered to have a particular meaning, which itself is not immediately present.[12]

In terms of coordinations of actions: naive objects are not entirely shared with others. Critical objects are a solution for this, as a way to coordinate the divergence between two persons' non-identical naive objects (cf. 6.3).

When people coordinate their behaviours in such a way that critical objects are brought forth, they may disagree about the essences, patterns or meanings that seem to underlie the naive objects that constitute its contents. Remaining conflicts consist of not being able to agree on the new form, or to bring it forth, e.g. the way in which a particular text is to be understood.

Conversely, to the extent that naive objects are (again) shared unproblematically, the use of critical objects disappears. No underlying pattern or meaning needs to be sought any more. Examples: the sounds a person makes are no longer sought to be interpreted as language, but just taken as they 'are'; the laboratory animals' behaviours are no longer interpreted at all in terms of pain or other experiences, but they are only observed in terms of the physiological phenomena about which the experiment is concerned.

6.8.

The critical observer may start to perceive the critical object's form (instead of merely conceiving it) as a naive object (cf. 6.6), when he succeeds in incorporating his already present naive objects as things *from which* the critical object can be observed. The naive objects (the critical object's contents[13]) then disappear as such, and become invisible, like the experiences by means of which this naive object could be observed (cf. 6.1) were also invisible. Maturana describes this in terms of an obscuration of the (coordinated actions that constitute the) earlier distinctions by the later distinctions (1987, pp. 359-360, 362).

New patterns between already present naive objects can be found again and again, and they become new naive objects by incorporating them into comprehensive entities. This can continue infinitely, and in fact this is what science is often concerned with: finding new phenomena or objects that are basic to, or generalizations of, already known phenomena or objects.

As an example of this we mention Von Foerster's experiments in training subjects to perceive 4-dimensional dice (tesseracts). It was through their learning to coordinate particular manipulations with buttons, which were connected to computer steered pictures, that these subjects were able to see the tesseract. In our terms: the tesseract became a naive object by incorporating the various pictorial changes related to the various button manipulations (cf. Von Foerster, 1985, pp. 74 ff.[14]). Likewise, good mathematicians are known for their skill to perceive the issues of their interest as 'really existing' things, and not merely as symbols between which formal rules apply. Then newly inferred aspects of these mathematical 'naive objects' may first be conceived only as the form of a critical object, and later become perceivable as naive objects by incorporating (by knowing them from) some of the already present naive objects.

Another example is mentioned by Shotter (1987): 'Castle Wolfenstein', a video game involving an intricately built imaginary edifice. Though Shotter presents it as an example of the construction of what he calls 'imaginary' objects, I tend to believe that this Castle Wolfenstein is as real as any other castle, but only as soon as one gets involved in the game and becomes able to find one's way through it, that is: as soon as one is able to perceive the castle *from* its constituent naive objects (i.e. the features of the video game). Then it passes from being a critical object to being a naive object, and it starts to be a comprehensive entity that is as really existing, like a piece of music that is being performed. Then its quality of video game moves to the background. It becomes also a consensual object for those who are involved with the game, just as the Dow Jones Index has become a real entity (and not just something imaginary) for those involved in the games of stock market transactions. To the extent that a player does not succeed in integrating its element, Castle Wolfenstein remains a critical object, a thought construct that one may definitely call with Shotter an imaginary entity. To the extent that this is the case, a comparison with a 'real' castle is made by means of a comparison of its components (bricks versus video-images). Only when considered at the level of its components (the naive objects), Castle Wolfenstein is an imaginary (that is, critical) object that can only be conceived, not perceived.

6.9.

Such a process of generating objects is one that continues and refines a particular already existing world. The relations between the new and the old objects are established in language. Anomic phenomena are observed and studied until they fit in with current theories, or until they convince people of the need for reformulation of current ideas. This holds both for scientific and common sense thinking. The new objects that come thus into existence arise at a particular place in a network of relations with other objects. Though they do arise due to second order distinctions, the first order distinctions that are dealt with as tokens for them, are themselves already second order distinctions (viz. those by which the (previous) naive objects were brought forth). These latter second order distinctions are now treated in turn as quasi-objects, as first order distinctions that can be (second order) distinguished (viz. those by which the 'meaning' or the 'essence' of the already existing naive objects are brought forth as new naive objects).

Thus, chains arise, each element in the chain being a naive object *from which* a new naive object can be observed, by incorporating it. An example is the mathematical objects mentioned in 6.8. Another example is the phenomenon of intersubjective consensus being re-evaluated as a true objectivity. Then two persons who from their own personal appreciations agree on the quality of a particular wine may conclude that the wine 'in fact' *is* a good one.

Notice that we have two modes of expressing the same phenomenon:

a) knowing a particular object *from* another naive object,
b) making a second order distinction of a first order distinction (which in its turn was itself a second order distinction).

According to expression mode a) a gestalt is perceivable as the new naive object, and the original naive objects become its elements that are no longer individually visible. Form and contents relate as new to old naive objects.

According to expression mode b) a second order distinction is taken (by a new second order distinction) as a new first order distinction. As a result, the previous first order distinction is no longer (second order) distinguished.

Combining these two expression modes, we may state that a naive object's form becomes perceivable as soon as the original first order distinctions are abstracted from, that is, as soon as they are no longer (second order) distinguished. It is as if the coming into existence of the naive object's form replaces the use of first order distinctions, and vice versa!

6.10.

We may distinguish two classes of naive objects: those that are new (to be called *original naive objects*), and those that are brought forth as a meaning or essence of an already existing naive object (to be called *derived naive objects).* Psychotherapy becomes useful in helping people bring forth original naive objects, which are principally based only upon first order distinctions (experiences). A person may start to construct derived naive objects wherever he can, led by the possibilities his language allows him. Then this person may alienate from his own original experiences, and these experiences are no longer used as a basis upon which second order distinctions are made, but rather are preceding second order distinctions rearranged and used as the new items to be coordinated (as in 6.9). Such a person may experience relief if he is enabled to make a new start of constructing objects out of experiences, instead of out of previous objects.

6.11.

A particular case of a chain of naive objects takes place when a loop in the chain occurs. Then two interacting persons may start to bring forth two incompatible descriptions of their interaction (cf. Watzlawick, Beavin & Jackson, 1967). The first order distinctions made by person A (the wife) is perceived by person B (the husband) as 'nagging', whereas his own first order distinctions are perceived by A as 'withdrawing'. Neither person perceives his own first order distinctions. Systemic therapies that use paradoxical injunctions of the variety 'keep nagging and keep withdrawing' are effective to the extent that each person is enabled to perceive his own previous first order distinctions, so that eventually they can develop a shared domain of naive objects.

7. *Mind: constructed or ineffable*

7.1.

'Internal' objects, such as experiences, representations, thoughts, emotions, intentions, dispositions, states, traits, etc., are usually assigned to a person as properties of his mental life that are situated inside the person. When dealt with in this way, they are of a constructed quality. Such an assignment of experiences to a person is one way to settle conflicts. The 'internal objects' that are assigned to a person are either critical objects or naive objects.

Like other objects, internal objects are a way to coordinate the divergence between two persons' objects. ("I am seeing the house as big and you are seeing the house as small; how is this possible ?; oh I see, we have different personal criteria for 'big' and 'small'".) Then the internal object that is created thus serves as a common point of reference to mitigate potentially conflictuous interactions. Thus sometimes in a psychotherapy, the patient is assigned a particular 'distorted' representation of reality. Then the therapy is aiming at a replacement of this 'distorted' representation by a more 'realistic' one. Especially cognitive behavioural therapies use this kind of thinking. Each interactant is allowed his own 'representation' or 'experience' of 'the' reality, that takes into account their 'personal bias factor'.

7.2.

It is by seeking answers to questions like "why does he disagree with me and consider this (objectively) safe situation as frightening ?" that people come to construct the internal representation of some person as something inside him. In constructing these internal representations and communicating them, we require from this person:

either

a) to reinterpret his own behaviours as indicators (predicates) of a particular mental state (substrate);

or

b) to reinterpret a particular meaning that he attaches to something as his 'subjective' predicate assigned to some 'objective' entity.

This is the basis of the social construction of 'experiences' and other 'mental' properties as properties that are 'in' the person who 'has' them.[15] In particular the psychoanalytic and client centred ways of psychotherapy often make extensive use of these internal objects and assume that they are 'there'.

In regard of the constructed quality of mental properties, Shotter (e.g. 1987) qualifies those entities as 'imaginary', that "...only 'exist' (*subsist* might be a better term) 'in' the activities *between* people".

Clearly, neither these imaginary 'subjective' entities, nor any other critical or naive objects are equal to the experiences that we equated above with first order distinctions.[16] For a critical observer who conceives a critical object's form does not see his own experiences (first order distinctions) as themselves naive objects, and neither does the observer who succeeds in perceiving this form as a naive object. Whatever we see as experiences, exists as constructed objects (cf. 7.1), not as actions that constitute first order distinctions. These are obscured by succeeding distinctions (cf. 6.8).

7.3.

A question that is of interest to us here, is why experiences as first order distinctions are so different from experiences as constructs (either naive or critical objects). The former resist objectification into 'real' things, whereas the latter, like Castle Wolfenstein or Von Foerster's tesseracts, seem to acquire the status of 'real' after more or less effort.

Two entirely different ways of knowing are the case. In order to come to perceive an 'internal object' as a naive object, it is necessary to incorporate the already present naive objects (that are usually particular behaviours of the observed person (cf. note 13), of which this internal object was originally conceived to be a basic pattern or form. As a result, we come to perceive the internal object *from* these more primary naive objects, and the internal object passes from a critical status to a naive one. This type of transition has been discussed above (cf. 6.6, 6.8). It is a transition in which a new naive object is brought forth.

On the other hand, access to first order distinctions (experiences) would require an entirely different transition, namely the kind of transition that occurs when a second order consensual distinction is no longer taking place and only a (smooth) first order coordination of actions remains (cf. 6.7.ii)). Access to first order distinctions requires a transition other than the bringing forth of objects (naive or critical), indeed its opposite: the very *cessation* of a naive object being perceivable. No incorporation of anything is required; to the contrary, the effect of any act of incorporation has to be made undone.

7.4.

Contrary to a Sartrean approach of understanding the unreflected consciousness as a nothingness, we may submit Merleau-Ponty's idea of a 'negative philosophy' (cf. Merleau-Ponty, 1964, p. 233) as a richer approach. Here unreflected consciousness ('the Flesh' (cf. 3.2), also called 'Being') is not so much put into the focus as an object of investigation. Rather our very (failing) efforts to get a glimpse of it are considered the only possible road to intuit it. This makes Madison state that philosophy's 'negativity' according to Merleau-Ponty does not so much concern its ".... "object" - it is not a philosophy of the subject or of Being conceived as *nothingness* - but rather its "methodology". Merleau-Ponty's ontology is a "negative philosophy" solely in the sense that for it the ultimate criterion of evidence is not, as Husserl insists, that the "object" of philosophical knowledge be present "in person", "in flesh and blood"; it is "negative" because it is not a "phenomenological positivism" which culminates in an eidetic intuition, in a *seeing*. (....) Consciousness can think Being or the flesh only as that which resists and hides itself from consciousness, and as latency and transcendence as such.

(....) This philosophy of Being worked out in the form of a negative philosophy does indeed therefore, by its method, resemble a "negative theology", because for such a theology the highest knowledge of God, the ultimate Transcendence, can be obtained only when everything has been done to know what God is and when one thus comes to know that one knows nothing of him; for it is then that one comes to know God, in his very proximity, as the totally other. (...) consciousness cannot think Being except by thinking its own blindness, what it cannot see because it is that which makes it see." (Madison, 1981, pp. 195/6).

Why are these remarks about Merleau-Ponty's notion of 'negative philosophy' for us of any use ? It is because they demonstrate a transition from the domain of second order consensual distinctions, into the pre-linguistic domain of first order consensual distinctions. It is by our (failing) attempts to objectify (that is: make perceivable as naive objects) our own first order consensual distinctions that we manage to leave the domain of second order distinctions. For when we try (by second order distinctions) to perceive (and therewith objectify) our first order distinctions, we attempt to take them as tokens, not for some newly emerging naive object, but as tokens for themselves!

7.5.

Then to the extent that we might believe to succeed in such an operation of objectification, the objects thus obtained are no longer the first order distinctions as we used them in order to perceive at all. At best the naive objects that we arrive at are some frozen images of them. In fact there are no second order distinctions possible that can fulfil the task of taking a particular first order distinction as a token for itself! For if there would be one particular second order distinction that could do this, how could it both a) use this self-referential first order distinction as a token and b) perceive it as an object ?

It is by conceiving this impossibility, and by noting the insufficiency of the naive objects that we are able to perceive, that we come to intuit this blind spot of our own reflexive possibilities.[17] We do not come to see it, but only to experience its presence as something beyond our observational capacities. In fact, it is not a single location, like the retina's blind spot, but rather a whole realm of unreflected and pre-reflective experiences. It is precisely here that language is no longer functional and even impedes a proper access.

7.6.

We are now able to formulate an issue that has a certain relevance for the psychotherapeutic practice. In section 2.3 we discussed the idea of an observer who sees a cat chasing a mouse. We called the mouse the quasi-object and the cat the quasi-observer. Whenever a person ('therapist') observes another person ('patient') and conceives (as a critical object that this patient is oriented towards) something which he (therapist) can observe ('the patient seems to be afraid of driving his car'), then he treats the patient as a quasi-observer oriented towards a quasi-object. Likewise, these roles of 'patient' and 'therapist' are condensable into one person who treats himself as a quasi-observer!

Psychological interpretations, made either by a therapist about a patient, or by a person about himself, are unfruitful to the extent that the patient is treated as a quasi-observer. Instead, an interpretation (literally) makes sense to the patient, to the extent that the therapist is able to (second order) distinguish the patient's (first order) distinctions! But in order to do so, the therapist needs to share these first order distinctions with the patient. In the next section we will spend some words on this.

8. *In conclusion: the one-sided boundary of a closed system*

8.1.

When we introduced the idea of autopoietic systems as organizationally closed (see 1.3), it was noted that a description of such a system would have to take into account that its states and its state transitions are of the same type. We also noted that for an observer of such a system a particular observational problem would arise when he tries to describe such a system's states. We are now able to be more specific about this.

In an arbitrary system (that does not have organizational closure), the state transitions develop in a trajectory that is triggered by the particular interactions this system undergoes. The interactions with an observer interested in the system's internal states, are no exception to this. For that reason, attempts to measure a system's states to some extent always lead to some degree of uncertainty (cf. Maturana, 1987, p. 343) about its states. This uncertainty may remain relatively small, but even so it may lead to errors in the prediction of future behaviours of the system.

The structural coupling that the system develops with the observer consists of state transitions in both (cf. 4.1). These state transitions take place as both interactants perturb each other's structures during the course of the interaction that is taking place between observer and system.

Precisely these state transitions that take place at the interface between observer (or his measurement device) and system, are not perceivable. For as a structural coupling between system and observer, they constitute the first order consensual distinctions, which, as we have discussed (cf. 7.5), are invisible for an observer. For it was by using these as tokens (i.e. as tokens for *something else*) that the observer is able to bring forth the system's state as a naive object.

8.2.

However, distinctive for organizationally closed systems is that their state transitions, which form the system's adaptive capacities (cf. 4.2), coincide with their states (cf. 1.3). Since the state transitions that take place as a consequence of the interaction with the observer cannot be perceived by the latter (see 8.1), as a consequence of this the system's states are as invisible to the observer as are its state transitions! An organizationally closed system, therefore, differs from other systems in that not only its state transitions, but also its states cannot be perceived by an observer. Any attempt to perceive its state transitions as states fails.

Whatever the observer will bring forth as the system's states, a coinciding with the state transitions will not be an essential feature of them, and for that reason they will not be the 'real'[18] states of the closed system. Whatever the observer will see then, it will not be the way in which the system compensates for the perturbations triggered in it during its interaction with the observer.

Any naive object that the observer brings forth from his first order distinctions will be a mere model[19] of the system (e.g. a computer simulation), that ignores the fact that in the system itself states and state transitions coincide. Any attempt to perceive these states as state transitions fails.

In other words: if we take the system's states as substrates and its state transitions as predicates to these substrates (cf. 1.5), then the system's closure entails predicates and substrates to be of the same type. The imperceptibility (not: inconceivability) of such a closure exists as the impossibility for an observer to use his first order distinctions as tokens for themselves. If a first order distinction is the form of that for which it is a token (cf. 6.1), then it cannot become visible as its own form.

The impossibility to observe the states of an organizationally closed system is primarily due to this closure; physical uncertainty relations (cf. 8.1) are only secondary. *The system's closure is as inaccessible to observation, as is the observer's own domain of first order distinctions (experiences).* If a person is to have any access to a closed system, it must be one that impedes his being an observer; it must be one that is beyond, or before, language (cf. 8.5).

8.3.

When Maturana and Varela speak of the 'domain of operations' of an organizationally closed system, it is precisely the domain in which the system's state transitions take place, that they consider inaccessible to the observer. Any observation the observer manages to perform is necessarily bound to the limitations of his perception, and therefore taking place in a descriptive domain that is not the domain in which the system itself operates.

Maturana maintains that "Given a closed system, inside and outside exist only for the observer who beholds it, not for the system." (Maturana, 1978, p. 41). Furthermore: "... a closed neuronal network cannot discriminate between internally and externally triggered changes in relative neuronal activity." (Maturana, 1978, p. 46); and: "For a living system in its operation as a closed system there is no inside or outside, it has no way of making the distinction. Yet, in language such a distinction arises ..." (Maturana, 1987, p. 364; cf. also Maturana, 1988, p. 57). Also Varela (1979, p. 243) maintains that it is the observer who describes the state changes of the autopoietic system as either internally or externally originated.[20]

We are led to the idea of understanding the autopoietic system in terms of its inside as demarcated from its outside. That what is inside is inaccessible to the observer: the domain of states and state transitions. That what is outside is accessible to the observer: the domain of behaviours and interactions of the system.

8.4.

The boundary of the autopoietic system that separates its inside from its outside, therefore, exists in the observer's domain of descriptions. The observer who speaks about the system's 'inside' behaves as a critical observer: he cannot perceive this inside (states and state transitions) in the way he can perceive the system's outside (behaviours); nor can he come to know this inside naively *from* the outside, without falling into the trap of constructing a model of the system that ignores the coinciding of states and state transitions. The critical observer can conceive the system's inside, but not perceive it. This boundary is not a property of the system's operations. Its states and state transitions do not contribute to this boundary; they are beyond it.

Therefore, it seems reasonable to characterize the boundary that separates the system's inside from its outside as *one-sided*: from the system's outside there is a difference between its inside and its outside; from its inside, on the other hand, there is not. This one-sided boundary exists only from its outside, as a demarcation brought forth by the observer; it does not exist from its inside. If ever the observer is to enter into the system's inside, it will be by no longer bringing it forth as distinct from the system's outside; that is: it will be by no longer making second order distinctions at all. The system's inside is inaccessible not because of a transcendental status (cf. 1.5); it is inaccessible due to the blind spot that the observer's own first order distinctions constitute for his perception.

8.5.

We are now able to understand the movement that we described in 7.3 in terms of a 'transition opposite to the bringing forth of objects': it is a movement towards the inner side of this one-sided boundary: from outside to inside. Crossing this boundary is only possible by leaving language, that is: by no longer distinguishing the boundary's outer side from its inner side. It is at its inner side that consensual (first order) distinctions allow the interacting individuals to have direct access to each others experiences. This is the 'natural empathy' that people are at times said to possess, which does not consist of a kind of 'reasoning by analogy', as is often argued by those who believe in the non-constructed status of what we have called 'internal objects'.

Instead, this natural empathy exists as a domain of first order consensual distinctions between interacting individuals. The shared experiences are accessible to both interactants. The 'inside' of the one is not distinguished from the 'inside' of the other: no 'insides' are distinguished at all! They both inhabit the same pre-reflexive world, and both the system's closure and the observer's own first order distinctions belong to it (cf. 8.2). Only when second order distinctions occur, do persons emerge to one another as naive objects to which, during the courses of their bargaining processes, an 'inner world' may be assigned.

8.6.

The task for an empirical psychology is not so much to assist in the bringing forth of all possible internal objects. Rather, it should study these very processes of bringing forth, as well as the reverse processes, by which the internal objects disappear. The notion of a one-sided boundary is helpful in this respect as a metaphor: entering into an organizationally closed system is not so much a matter of 'opening' it in order to observe it, but rather a matter of loosing and loosening one's observational/linguistic mode of functioning in order to 'un-distinguish' that what before existed as an observable 'objective' reality.

Notes

1. We are reminded of the original meaning of the verb 'explain', i.e. 'flatten' or 'even out'. By such an act something becomes stratified in a way it was not before.

2. Cf. Kauffman & Varela, 1980, p. 205: the 'act of distinguishing' as indistinguishable from the 'distinction'.

3. We are reminded of Von Uexküll's (1909) term 'Umwelt' (cf. Maturana, 1980,p. 51).

4. Cf. Maturana 1978, p. 48; 1987, p. 360.

5. Indeed, we may wonder whether this entails that the difference between man and animal is not the use of language, or whether it means that some animals have capacities to behave humanlike.

6. This does not mean that a critical observer is defined as having an awareness of predicates and substrates: like the terms 'form' and 'contents' they belong to the domain of a meta-observer.

7. Cf. Merleau-Ponty: "Il y a un *Wesen* du rouge, qui n'est pas le *Wesen* du vert; mais c'est un *Wesen* qui par principe n'est accessible qu'à travers le voir, et est accessible dès que le voir est donné, n'a dès lors plus besoin d'être *pensé*: voir, c'est cette sorte de pensée qui n'a pas besoin de penser pour posséder le *Wesen* ..." (1964, p. 301).

8. Cf. Polanyi's, 1966 use of this term. Furthermore, Polanyi uses the term 'particular' for what we have called a component, that, together with other naive objects, constitutes the contents of the critical object.

9. The activity that transforms a naive object into a token for a newly emerging naive object, is one that Polanyi (e.g. 1966; see also Shotter, 1987) calls 'knowing from'. It is a mode of acting by which some object is used as a tool in perceiving the newly emerging naive object. It is by looking through these tools i.e., by *'knowing from'* them, that the naive objects arise.

10. We are reminded of Maturana term 'simple unity' (1987 p. 333, that exists in a space defined by its properties, and not by any other additional or derived properties (such as the properties of its components). The reason for not following Maturana's terminology here is that we are not so much speaking about conceivable unities, but rather about perceivable objects.

11. We are reminded of Maturana's notion of 'composite unity' (1987, p. 333), the components of which we are presenting here in terms of synthetic properties.

12. This according to case B (see 6.4). Case A can be understood in a similar way: the naive objects available share a form that is their essence. This form is some underlying pattern (cf. Bateson's (1979) notion of 'the pattern that connects') that is recognizable in its various appearances that are present as the visible naive objects (i.e. the predicates).

13. These may consist of several naive objects between which a particular regularity is to be found, or one particular object that over time appears in a variety of ways.

14. Cf. also Von Foerster's statement "nothing exists, unless you can grasp it" (Lecture given at Bergamo conference "Evolution and Cognition", 1990).

15. In fact there is no compelling reason why these 'internal' objects should be located inside a person. Gods, ghosts, demons, spirits, satyrs or nymphs, as they occur in most civilizations, are no less real than the mental objects of our culture.

16. The reader is reminded of classical distinctions between consciousness as pre-reflexive (cf. what we have called 'first order distinctions') versus consciousness as reflected upon (cf. what we have called 'internal objects'). Each attempt to observe, measure or 'operationalize' the former will yield the latter.

17. *Ce qu*'elle [la conscience] ne voit pas, c'est ce qui en elle prépare la vision du reste (comme la rétine est aveugle au point d'où se répandent en elle les fibres qui permettront la vision). (Merleau-Ponty, 1964, p. 301).

18. 'Real': not in the sense of transcendental, but, to the contrary, in the sense of taking place closely to the observer, at his own interaction interface. Indeed, this very proximity to the observer impedes his perception of it.

19. Cf.: "We as observers can project all cellular processes upon a system of three orthogonal coordinates, and legitimately say, as valid in the projection, that specification is mostly produced by nucleic acids, constitution by proteins, and order (regulation) by metabolites. The autopoietic space, however, is curved and closed in the sense that it is entirely specified by itself, and such a projection represents our cognitive relation with it, but does not reproduce it." (Maturana & Varela, 1980, p. 92).

20. Cf. the notion that the autopoietic system realizes its own boundaries (see e.g. Maturana, 1975, p. 317; Maturana & Varela, 1980, p. 81). The system's boundaries are considered the outcome of its operations, not as the result of the observer's definitions. We are reminded of Anselm's definition of God as the greatest that can be conceived ("id quo maius cogitari nequit", 'IQM'). Anselm argues that this definition entails God's existence: IQM cannot be conceived but as actually existing, or it would not be the greatest. Were it non-existent, then IQM could be conceived as being even greater than something that exists merely conceptually (cf. de Rijk, 1981, pp. 149ff.). Likewise, the definition of an autopoietic system entails that it realizes its own boundaries in a way that is *not* defined by an observer! Therefore, were the observer to define the system's boundaries, he would treat the system as allopoietic. Indeed, Zolo's criticism that the ideas presented by Maturana & Varela (1980) resemble scholastic issues is not too much beside the mark (Zolo, 1986, p. 145). (However, apparently unlike Zolo, I tend to believe that many issues of medieval philosophy are still of great importance.)

References

Bateson, G., 1979, *Mind and nature. A necessary unity*, Dutton, New York.

Ciompi, L., 1982, *Affektlogik. Ueber die Struktur der Psyche und ihre Entwicklung. Ein Beitrag zur Schizophrenieforschung*, Klett-Cotta, Stuttgart.

Kauffman, L.H.; Varela, F.J., 1980, Form dynamics, *J. Social Biol. Structures*, vol. 3, pp. 171-206.

Madison, G.B., 1981, *A search for the limits of consciousness. The phenomenology of Merleau-Ponty*, Ohio U.P., Athens, Ohio.

Maturana, H.R., 1975, The organization of the living: a theory of the living organization, *Int. J. Man-Machine Studies*, vol. 7, pp. 313-332.

Maturana, H.R., 1978, Biology of language: the epistemology of reality, in *Psychology and biology of language and thought. Essays in honor of Eric Lenneberg: conference*, G.A. Miller and E. Lenneberg (eds.), Academic Press, New York.

Maturana, H.R.; Varela, F.J., 1980, *Autopoiesis and cognition. The realization of the living*, Reidel, Dordrecht/Boston.

Maturana, H.R., 1980, Autopoiesis: reproduction, heredity and evolution, in *Autopoiesis, dissipative structures and spontaneous social orders*, M. Zeleny (ed.), Westview, Boulder, Col.

Maturana, H.R., 1982, Repräsentation und Kommunikation, in *Erkennen: die Organisation und Verkörperung von Wirklichkeit. Ausgewählte Arbeiten zur biologischen Epistemologie*, H.R. Maturana, Vieweg, Braunschweig-Wiesbaden.

Maturana, H.R., 1983, What is it to see ?, *Arch. Biol. Med. Exp.*, vol. 16, pp. 255-269.

Maturana, H.R., 1987, The biological foundations of self-consciousness and the physical domain of existence, in *Physics of cognitive processes. Proceedings of the International Symposium on physics of cognitive processes*, E.R. Caianiello (ed.), Amalfi, Italy, 16-20 june 1986, World Scientific P., New Jersey.

Maturana, H.R., 1988, Reality: the search for objectivity or the quest for a compelling argument, *Irish Journal of Psychology*, vol. 9, no. 1, pp. 25-82.

Merleau-Ponty, M., 1964, *Le visible et l'invisible*, Gallimard, Paris.

Milner, P.M., 1971, *Physiological psychology*, Holt, Rinehart, Winston, London.

Minsky, M.L., 1967, *Computation. Finite and infinite machines*, Prentice-Hall, London.

Polanyi, M., 1966, *The tacit dimension*, Doubleday, New York.

De Rijk, L.M., 1981, *Middeleeuwse wijsbegeerte. Traditie en vernieuwing*, Van Gorcum, Assen.

Shotter, J., 1984, *Social accountability and selfhood*, Basil Blackwell, Oxford.

Shotter, J., 1987, Social accountability and the linguistic creation of mind: the unaccountability of computer conduct, in *Computers, mind and metaphors*, K. Plunkett, S.F. Larsen (eds.), Aarhus U.P., Aarhus/ Erlbaum, Englewood Cliffs, NJ.

Shotter, J., in press, A poetics of relational forms: the sociality of everyday social life, to appear in *Cultural dynamics*, special issue on 'Evolutionary models in the social sciences', T. Ingold (ed.), (page numbers from submitted draft version).

Varela, F.J., 1979, *Principles of biological autonomy*, North Holland, New York

Varela, F.J., 1983, l'Auto-organisation: de l'apparence au mécanisme, in P. Dumouchel and J.-P. Dupuy (eds.), *l'Auto-organisation: de la physique au politique*, Seuil, Paris.

Von Foerster, H., 1985, Kybernetik einer Erkenntnistheorie, in *Sicht und Einsicht*, H. von Foerster, Vieweg, Braunschweig.

Von Uexküll, J., 1909, *Umwelt und Innenwelt der Tiere*, Springer, Berlin.

Watzlawick, P.; Beavin, J.H.; Jackson, D.D., 1967, *Pragmatics of human communication*, Norton, New York.

Zolo, D., 1986, Autopoiesis. Critica di un paradigma conservatore, *MicroMega*, no. 1, pp. 129-173.

MECHANISTIC EXPLANATIONS AND STRUCTURE-DETERMINED SYSTEMS
MATURANA AND THE HUMAN SCIENCES

Paula Burghgraeve

Free University of Brussels (VUB)
Department of Philosophy of Science

Abstract

The notion of mechanistic explanation is given a very specific meaning by Maturana. Does it influence research in human sciences ? In this article an attempt is made to answer this question.

Introduction

When first confronted with Maturana's description of a living system, his general point of view made sense to me.[1] I did not especially try to know everything he said or wrote in detail. This is probably the reason why it took me some years before I decided to read the 1978 article 'Biology of Language: The Epistemology of Reality.'[2] I decided to read it, because it is claimed that this article is the real clue to Maturana's thinking. After reading it, my conclusion is that it is a very difficult article. I didn't come to formulate a personal view on it.

Two or three times I started to read it again. The more I looked at it, the more I became convinced that, if Maturana proves right in his approach of systems in general and living systems in particular, our outlook on human sciences has to change drastically. This impression is shared by other researchers and is generally expressed by stressing the fact that no point of view, theory, etc., on man and society can objectively be preferred to any other point of view, theory, etc.

So, what I am writing down now, is an attempt to follow Maturana's thinking with the intention to answer the question: does Maturana's point of view make a difference to further research within the field of human sciences ?

First I thought I would find the necessary arguments when going deeper into the ideas he formulates concerning living systems. To my own surprise I didn't need to go into particulars with regard to autopoietic systems. A sufficient reason already comes to the fore when looking at his description of scientific research.

G. Van de Vijver (ed.), New Perspectives on Cybernetics, 207–217.

1. The origin of characteristics

Being involved as I was with AI, I decided to start looking at the paper from the angle of functionalism. Functionalism itself is not an easy concept. It has several slightly different meanings, which are too complex to take into account within the context of this article. I use this approach only as a starting point and a reference. Hence, I start from a very general definition of functionalism as a specific way to study systems.[3]

A functional approach to study systems implies the analysis of the system into parts, the description of the specific relations between the components and the description of the interaction or processes taking place between them, with the intention to explain why the system behaves as it does. Usually research based on a functional approach includes the specification of the relations between certain parts of the system and the environment of the system.

If we take this general definition as a starting point, we can use Maturana's approach to scientific explanations and characterize a functional system description as an attempt to give a mechanistic explanation of a particular system.

At first I thought this characterization of a functional description to be quite correct, since Maturana states that it is only in mechanistic explanations that the observed system is treated as a composite unity. But I became less sure of this when trying to proceed with the interpretation, using my definition of functionalism as a leading idea.

Maturana states that he distinguishes between two standard scientific explanations. On the one hand he introduces the vitalistic explanation, which explains the characteristics of a system by stating that they are found among the properties of a component of the system. On the other hand, the mechanistic explanation explains characteristics and properties of the system as the result of the specific interaction of the components of a system and not as characteristics of components as such.

Classifying the research as mechanistic, means that it is necessary and sufficient to state that the system under study only behaves as it does because all the components contribute to it. Therefore, a mechanistic explanation necessarily needs a decomposition of the system into components and their interrelations. A vitalistic explanation doesn't need this decomposition, because the specific interaction between the specific components doesn't play any part in the description of the characteristics of the system.

Can we state that a functional approach which tries to explain why a system behaves in a particular way, always leads to a mechanistic explanation ? Following Maturana, I am inclined to give a negative answer to this question. We could certainly imagine a model of the brain, starting from the assumption that the property of humans to remember is the sole result of the property of a memory-unity, that the property of thinking is the sole result of a thinking engine such as a computation-unity, etc. And this way of looking at human cognitive abilities is not as far-fetched as one would think, because this has been a fairly common way of speaking metaphorically about human cognitive abilities since the beginning of computer age.

The example is rather an obvious one and I use it to make it clear that one could easily think that one is explaining something mechanically, but that this doesn't imply that this seemingly mechanical way to explain a phenomenon is what Maturana means when he is talking about a mechanistic explanation. The obvious conclusion is that the decomposition in parts is not a guarantee for a mechanistic explanation, but that it is vital that no characteristic of the complete unity should be explained without reference to the interrelations of the parts. Hence, not all functional analyses resolve into a mechanistic explanation.

Assuming that I understand the difference between vitalistic and mechanistic explanations correctly, I could make a risky generalization and state that from Maturana's point of view, every methodological stand, up to this moment, proposes obscure views as to the choice of the origin of characteristics. In other words: researchers don't seem to know exactly why and when an explanation is a mechanistic one and why and when an explanation is a vitalistic one.

The obvious question now is why they didn't have a clear view of the difference between those two forms of explanation.

2. *Making distinctions*

A unity comes into being through operations of distinction. Those operations assign properties to something. Making a distinction, and as a consequence assigning properties and hence seeing something as a unity, depends on the observer. Being able to see a 'thing' is the result of the possibility of the observer to be aware of a difference between 'it' and another 'thing'.

Two crucial notions emerge here: 'making distinctions' and 'characteristics'. They seem crucial to me because they play a part in the distinction between mechanistic and vitalistic explanations. So let us examine them more closely.

When I start to analyze something functionally, I want to find the reasons why, or in other words, explain why a system behaves as it does. This means that when I make the distinction between my daughter and my cat, I classify them within two different classes, not because I first dissected them, but because I made a distinction as to their behaviour.

But I could easily imagine other ways to make a distinction between my daughter and my cat. I look at them and without waiting for them to act I see that they definitely haven't the same physiognomy. Because I am a human being with a particular education, I can find all sorts of other ways to distinguish between cats and human beings. Even when only taking into account the distinctions based on behaviour and those based on physiognomy as a starting point, I could easily realize that not every possible distinction is suitable for a mechanistic approach.

3. Making distinctions leading to mechanistic explanations

Maturana must have been aware of this. Hence we have to look for further specifications of the mechanistic explanation. He offers some help by the way he entitles the second part of his article: 'Operational Characteristics of a Mechanic explanation'.

Organization, structure and space are the three main notions which can shed a light upon this problem. I start with the notion of space, because it seems to me that this notion plays a crucial part in understanding the particular choice of the operations of distinction to make a mechanistic explanation possible.

"Space", he says, "is the domain of all the possible interactions of a collection of unities (simple, or composite that interact as unities) that the properties of these unities establish by specifying its dimensions." And further: "Once a unity is defined, a space is specified." (Maturana, 1978, p. 33).

The moment I decide to point at something as a unity, I can specify its space. It becomes immediately clear that every identification gives rise to two particular spaces, because every unity can be viewed as simple or composite.

Let us try to see what this statement amounts to by alternately looking at a human being as a simple and as a composite unity. As a simple unity, the space of a human being definitely lies at the outer side of his skin. I characterize him, for example, by taking his social behaviour into account, by noticing that he has neither feathers, nor fur etc. As a composite unity, the space of a human being definitely lies at the inner side of his skin. All what is going on inside can be taken as characteristics to establish the human being as a composite unity.

In addition to this central dichotomy, we have to take into account the actual possibilities of an observer to observe. These can change and this change can influence the spaces he is establishing.

Let us now look at the notion of 'organization'. This is one of the main notions playing a part in mechanistic explanations. It is the organization of something that establishes it as a composite unity and organization refers to the specific instrumental participation of components. The composite unity, defined by its internal organization, establishes a subject-dependent mechanistic explanation.

To recapitulate, at the very outset I identify my daughter as a simple unity by establishing an outer skin-space. Depending upon my choice as an observer, I can, for example, use characteristics - being properties resulting from operations of distinctions - that make me look at her as my daughter and no one else, or generally as a human being, or as a mammal, etc., but only if I start to search for inner skin characteristics do I look at her as a composite unity and only then the conditions are fulfilled to start the construction of a mechanistic explanation.

If, up to this point, I understood Maturana correctly, I can say that when choosing, as an observer, a particular set of intellectual tasks as the outer skin characteristics of human beings, it is possible to identify human beings and computers as simple unities belonging to the same class of entities. It seems to me that this is exactly what the Turing test amounts to. But at the same time, this identification, based on the specifically chosen outer skin characteristics, excludes a mechanistic explanation of the identified entities. To identify them from a mechanical point of view, I have to specify inner skin components and relations between components which allow for an organic definition of both entities.

To keep the argument clear, I will give the description of the third main notion later on in the text. First I want to come back to my functional approach. When I say that I want to explain how a unity behaves by splitting it up into parts, I can imagine an objection from someone following Maturana. He or she immediately will attract our attention to the fact that outer skin characteristics and inner skin characteristics are two quite different things. He or she may easily wonder whether it is allowed to say that one explains outer skin characteristics by establishing connections with explanations of inner skin characteristics. The behaviour of the entity as a simple entity emerges from all that is happening inside, but cannot be reduced to a description of a particular inside activity.

4. Making differences between entities from a mechanistic point of view

A mechanistic approach explains what is going on in the inner skin action space of a unity. To decide that something is no longer a member of a particular entity class from a mechanistic point of view, is to state that there is an internal difference between this thing and the members of the entity class.

If I go back to functionalism and artificial intelligence, I can state that a computer-representation functionalism is undoubtedly a mechanistic approach of the thing called computer, but can only be a mechanistic approach of a human brain if they are both - computer and brain - internally identical. Knowing then how Maturana defines organization, can we claim them to be organizationally identical ?

I deliberately didn't go into the notion of 'structure' in the previous section, because I think that this notion can shed some light upon an additional specification of a mechanistic explanation.

Organization as well as structure are concepts used by Maturana to say something about the way of existing of unities, but it is only the concept of organization that plays a part in the identification of a class of composite unities.

Let us restrict ourselves to the notion of organization, and see whether we can give an account of the brain organization that is equivalent to the one we use to identify a computer. We only have to turn to an arbitrary textbook on computer science and we shall always find the same description of the organization of the computer. The parts of the composite unity 'computer' amount to: an arithmetic unity, a control unity, a memory or storage unity, an input- output unity and their specific interactions. And without the slightest chance of being corrected, we can state that the mind can be viewed in the same way.

This changes if we take the notion of 'structure' into account. The structure of a composite unity refers to 'the actual components and the actual relations that these must satisfy in their participation in the constitution of a given composite unity' (Maturana, 1978, p. 32).

To refer to the actual components and actual relations of a computer, and hence to refer to the structure of a computer before and after installing new software, would mean that we are talking of the same computer with a different structure at different moments in time.

We can use an even better way of presenting this example to stress the difference between organization and structure. The organization of a computer in terms of a Turing machine amounts to the specification of a set of conditionals in the form of a machine table. The structure of this particular computer consists of the actual values we get after, let us say 100 steps, given the aforementioned conditionals. In this formalism, the organizational identification of a computer and the mind could be stated from the moment they have one and the same machine table. To refer to their structure would mean to look at the actual values their conditionals generated after a certain period of time.

In the above 'examples' I deliberately used 'mind' instead of 'brain', because if I start constructing a definition of the organization from a real human brain, it becomes quite difficult to insist on identical organizational definitions of brain and computer.

5. The identification of structure-determined systems

Only because he uses this particular description of mechanistic explanations, can Maturana come to the determination of a class of entities as structure-determined. Structure-determined systems either undergo changes of state 'determining the particular structure at a particular moment', or they undergo changes leading to a loss of identity.

There would be nothing very spectacular in the specification of this particular class of entities, if the introduction wasn't accompanied by some penetrating remarks about how changes come about in structure-determined systems.

Perturbing interactions

In the first place, there is the striking remark that changes of state of a structure-determined system can be triggered by perturbing interactions, but that they are not determined by the properties of the entity producing the perturbing interaction. It is very important to understand this perfectly.

Taking again the computer as an example may be quite helpful. Using a function key implies the execution of a particular sequence of instructions. The perturbing action is, in this case, the simple use of a particular function key. When taking the same finger, and using the same, but redefined function key, it becomes quite clear that the particular interaction (using the key), coming from a particular entity (person), with specific properties, doesn't play a part in the specific developments of the internal structural change. Only the actual structure determines what's going to happen and the perturbing entity has nothing to do with the actual change. It is only because the computer is such a 'simple' device, that we can predict the change following the use of a function key.

Applied to a human being this would for instance mean that to say that someone is influenced by someone else, only refers to the fact that the last person is acting as a triggerer of structural change with regard to the first one, but that the triggerer isn't sure that he can be a triggerer, nor what will be the result of being the triggerer.

If we say of something that it is structure-determined, this necessarily means that we look at the system as a composite entity. We know though, that to look at something as simple or as composite depends on the choice of the observer. Knowing that a system is structure-determined doesn't prevent the observer to see overall characteristics of the entity. One specific form of characteristic of the entity as a simple entity can be a susceptibility for certain forms of perturbing interaction. This results in particular structural changes only determined by the organization and the previous structure of the system. This can lead to a change in characteristics of the entity, looked upon as a simple entity, etc.

Instructions

The remark about the absence of influence coming from the properties of the perturbing entities has to be seen in connection with the remark about instructive interaction. Maturana says: "Systems that undergo instructive interactions cannot be analyzed by a scientific procedure... all instructable systems would adopt the same state under the same perturbations and would necessarily be indistinguishable to a standard observer." (Maturana, 1978, p. 34).

The notion of instruction has again a very specific meaning. Instruction is explicitly coupled to the idea that properties of a perturbing entity could specify the changes of state a system goes through. Those systems are, by definition, not structure determined. The fact that Maturana states that they cannot be analyzed scientifically, because they would be indistinguishable to the standard observer, caused me to have some problems at first. I only could interpret this remark by imagining a system that had escaped from the world of science fiction, that had the means to change itself perfectly and instantly into the perturbing entity. We shall see that the system involved needn't be so strange and that at least partial instructable systems form an important part of our world.

Living systems are autopoietic. Autopoietic systems are a particular subset of structure-determined systems. They are in fact autonomous entities and their autonomy is a direct consequence of their particular organization which allows the system to arrive at a continuous self-production of the necessary components and relations between components to maintain its organization. Because autopoietic systems form this particular subset of structure-determined systems, they certainly cannot be instructed. Their structural change has nothing to do with the properties of perturbing entities, but has everything to do with their own internal structural situation at a particular moment in their personal history.

Every outer skin property, or, with the words of Maturana, every characteristic of an autopoietic system seen as a simple entity, is the result of an ontologically completely determined change of structure. The autopoietic entity as a simple entity has no means to act upon the completely determined internal structure that is actually responsible for the actual domain of perturbations. It has no means, moreover, to control all the perturbing entities that are made possible by its structure.

6. Explanation in human sciences

We now switch to human beings as autopoietic systems. Instead of following Maturana through the concepts of structural coupling between structurally plastic organisms and without making use of the notion of consensual domain, we can ask ourselves if it is possible to analyze the characteristics of a human being as a simple entity from a mechanistic point of view.

Characteristics of it are for example the use of language, the production of artifacts, etc. This implies two things: on the one hand, physical traces such as buildings, books, pictures, paintings and the very specific outlook of those physical traces, on the other hand, shortlived behaviour that can be the perturbing factor to another structurally plastic organism.

To analyze the physical traces mechanically means that one tries to define the organization of the physical object. The organization of the objects makes it possible to classify them into categories such as books, computers, etc. But the organization doesn't account for the very particular shape the objects have. This very particular shape is the result of a transfer, of a partial instruction. The properties of the disturbing entity 'human being' at a particular moment of its existence — having a particular structure, which determines and individualizes its behaviour — are transferred to the entity under consideration. The entity is a frozen behaviour of the human being as a simple entity engraved on a particular physical entity. This is the only way I can imagine instructable systems in our particular world.

Maturana says that instructable entities cannot be analyzed scientifically because a standard observer wouldn't be able to distinguish those instructable systems, as they and the disturbing entity would be quite alike. But since the instructable entity consists of a frozen instruction, it must be possible to differentiate it. The content of a book doesn't change any more once it is written down. The entity is not an instructable one any more. And since the physical traces of frozen human behaviour are not submitted to structural change, this content cannot be viewed as a structure-determined entity. The book can be a perturbing agent for a human being, but its frozen behaviour cannot be perturbed by the act of reading.

Hence every example of a physical trace of human behaviour can be differentiated by an observer. Every physical trace can be viewed as a simple entity or as a composite entity. "Everything that is said about the physical trace, is said by an observer to another observer, who can also be himself or herself." (Maturana, 1978, p. 31). Hence, we can create a "closed cognitive domain in which all statements are, of necessity, subject dependent, valid only in the domain of interactions in which the standard observer exists and operates" (Maturana, 1978, p. 29). Consequently, if we accept that the properties of frozen human behaviour are to be explained by the relations of the components of the frozen behaviour and that they are not to be found among the properties of the components, we start producing a mechanistic explanation of this particular form of entities.

This argumentation cannot be repeated in the case of the short-lived behaviour, although the latter can play an important part in the change of autopoietic systems. Hence, part of the analytical requirements to make the appropriate mechanistic analysis, cannot be fulfilled.

Is there any possibility to use the scientific research of frozen structural states to predict and control autopoietic systems ? Yes and no, I should say. Yes, if it would be possible to produce a complete machine table for every second-order plastic structure and achieve complete control over every structure trajectory of every structure-determined system. No, because this isn't possible for the very good reason that, while trying to pin down such tables, we are, as second-order plastic structures, not only changing the state we are in, but we are also changing components involved in the determination of our own subsequent structural change. Every machine table could at best be a partially frozen witness of a past situation, and a starting point with an endlessly large amount of possible ways to change. The question would remain thereby: in what state are we now and what will be the following one ?

7. Conclusion

This conclusion cannot be a surprising one. But it makes one thing particularly clear: one cannot participate and observe at the same time. Or differently said: participating and observing are mutually exclusive ways of structural change. This is bound to have consequences for every theory of the human sciences that tries to explain and predict the actual behaviour of human beings. Prediction as a result of a mechanical analysis of actual psychological or sociological behaviour isn't a realistic goal from this point of view. If Maturana's approach is accepted, human sciences have to search for another use of their models and theories. Which goal shall we select and why on the basis of this insight ? And shall we not have to change our ways of observing ourselves drastically ?

Notes

1. F.G. Varela, H.R. Maturana and R. Uribe, 1974, Autopoiesis: The Organization of Living Systems, its Characterization and a Model, *BioSystems*, vol. 5, North-Holland Publishing Company, Amsterdam.

2. H.R. Maturana, 1978, Biology of Language: The Epistemology of Reality, in *Psychology and Biology of Language and Thought. Essays in Honor of Eric Lenneberg*, Academic Press.

3. A good overview and a critique of functionalism is available in N. Block (ed.), 1981, *Readings in the Philosophy of Psychology*, Methuen & Co, London.

SOCIOLOGICAL ISSUES

CORRESPONDENCE, CONSENSUS, COHERENCE AND THE RAPE OF DEMOCRACY*

Gordon Pask

University of Amsterdam
Faculty of Andragological and Educational Studies

Abstract

It is argued that a confusion among correspondence, consensus and coherence truth values is counterproductive, notably in the social field. Refinements of this rough categorization are discussed, as well as novel interpretations like dynamic coherence, akin to 'hermeneutic' truth regarded as 'the iterative refinement of meaning'. It is maintained that dynamic coherence is fundamental to the other forms of truth prior to them, especially in specific domains of interpretation. Societal implications are considered insofar as a democratic organization is intended to exist, notably the difference between 'unity' and 'uniformity'.

1. Introduction

There are, of course, many varieties of truth. For example, there is absolute truth, factual truth and analytic truth. This paper is concerned with only the major divisions; correspondence truth (often called veridical), consensual truth (majority votes and the like) and coherence truth, applicable to reasoned debate, agreement and agreement to disagree over some matter. In its dynamic form this is also known as 'hermeneutic truth', if, with Taylor, for instance, we intend 'the iterative refinement of meaning'.

A great deal of confusion appears to exist between these categories and the domains to which they apply, which is odd. Coherence truth and its dynamic version, hermeneutic truth, has a primacy in logic and its kinetic implementation is not only logically prior to the other types but carries the value, also, of existence. Further it is crucial in any understanding of such notions as 'organizational closure', in general, of P-Individuation and 'autopoiesis', the latter in biological domains.

G. Van de Vijver (ed.), New Perspectives on Cybernetics, 221–232.

Further, some forms of this common misperception lead to an often repeated rape of that lovely lady, democracy. The name 'democracy' means, so the dictionary says, 'government of the people by the people or their agreed executives', which, so frequently, is precisely what it does not mean, in practice. One, at least, of the main reasons for this degradation is that the 'coherence' of agreement after (often prolonged, often dialectic and even heated) debate is substituted by the simple majority logic of a 'consensus'. The matter becomes critical whenever some forced choice is ordained, for example, by voting, and when the voting process is open to manipulation as it usually is (and I do not mean falsification, as in rigging votes).

2. *Correspondence truth*

Correspondence truth, which is often given the forefront of the stage since it is the stuff of factuality, is, logically, subordinate to common agreement and coherence truth. Science, in particular, is said to depend upon it. But, in order to establish the correspondence truth of a statement (that the proposition, declaring an hypothesis, has veridical tenure, being in correspondence with some fact of evidence), there must, first of all, be agreement that the hypothesis is worth the testing and that an appropriate test has been or can be devised.

Really these are social, minimally psychological matters. They all involve debate, conversation, dialectic or, in the limit, diatribe. That is in the realm of coherence truth, as will be shown. It is not a matter of discovering serendipitous occurrences, which some call eoliths (stones that have the shape to function as spades). For, although I would be loath to dismiss either serendipity or eoliths, neither of them are the main substance of science nor the primary funds of commonplace wisdom.

Suppose, however, that an hypothesis, part of some theoretical body of intertwined hypotheses, has been agreed as being worth the test. Suppose also that means of testing, instruments and so on, have been agreed. If so, then it is either possible to devise a Popperian type, crucial kind of hypothesis, the falsification of which would lead one to discard some or all of the theory or else, more commonly, to conduct trials in order to affirm, by repetition, that the hypothesis *does* correspond in important aspects, to the evidence. Thereby we may proceed to formulate scientific laws, general laws, or practically useful principles (like the notion that a cup drops, if I let it fall) by dint of 'gravity' rather than happenstance (perhaps because 'gravity' has a repeatably determinable acceleration constant, known as g).

All that is excellent. But over and above the caveat about the ultimate dependence of correspondence truth upon various kinds of coherence or agreement, another caveat is needed. Correspondence truth is a rare treasure and when it is found it is, for the most part, restricted to specific domains, or possible worlds, or universes of discourse. That assertion in no way derates its value, which is immense, both practically and theoretically. On the other hand it stresses the prudence of taking care when overgeneralizing, for example, when using statistical methods, of which many are confined to binary reducible relations and still more to the knot-and-string progression of schoolboy mechanics in regions where assumptions of this kind are quite improper.

3. Other limits

Apart from these comments, some other, more persnickety cautions are in order. There is no such thing as the correspondence truth of a command or a question addressed from a participant A to some other participant B, to the obedience of A or B, or their replies or not to the questions. A or B may disobey, or reply by a further enquiry. Such statements as entreaties and requests follow along similar lines but with greater complication, due to their elaborately conditional nature.

Neither do metaphors designating analogies or the allegories of stories have, in general, correspondence truth values 'Yes' or 'No' unless they are, perhaps, observed histories (when the evidence stated depends upon the observer). Of course it is possible to make meta-statements about commands, questions, and the like (statements made by an observer about some event, the inherent logical character of which is not, as a result, necessarily given fresh meaning). For example, one can say that 'A commanded B to do C if certain conditions pertained at a certain moment, designated D'. If the observer responsible for this statement also asserts that 'it is true that A did command B to do C when D happened, that D happen, that B obeyed', then it is legitimate to say that the observer's meta-statement of 'command termination', equally of 'question acception and response' is true. But these addressed statements of fiat or interrogation do not, in themselves, have correspondence truth value.

By much the same token, most laws of science are literally meta-statements in an abstract analogical universe, relating laws of the same form in particular universes of hydraulics or electricity. Events in these particular universes may be similarly structured and, if so, the scientific law is an isomorphism, or one to one correspondence. To claim more for the law 'that it is reality' or that all universes are identically indexed and reducible to one kind is 'reductionism' approaching a neo-positivistic stance.

It thus appears that we must go beyond the tidy domain of correspondence (veridical, factual), truth when dealing with most matters of consequence in the social, psychological, and probably other spheres. In particular such excursions are needed when speaking of people, individuals and groups, capable of addressing edicts or requests or responding to them. Again, such excursions prove mandatory when bridging the many gaps between the different domains, worlds, or universes of discourse which are normally encountered. In fact, it seems right enough to assert that some journey beyond correspondence truths alone is needed whenever (which is usually) an observer is neither numinous nor omniscient.

Where does the journey lead ? Does it go to coherence, or to consensus, or to both. Are they always compatible with each other ?

4. Consensual truth

One kind of consensual truth is complete accord about some thing or event. More commonly, it is a matter of majority. The particulars of the majority logic employed in order to reach consensus and decide (whether or not such a thing or event is the case, or how to view it under given circumstances) are various. An outright majority has always a maximal consensual truth value but percentages (so may votes for, so many against out of a total, or out of those who did not abstain, or whatever) may be handled as a probability number or by giving authority to the greatest, with tricks for handling the deadlock, or by distributing authority over several factions.

It is worth looking at some of the assumptions that underlie any such assignment of a consensual truth. For this purpose, it is useful to suppose that vote casting over a specified interval, as at an election, is the means for determining the consensual value. The comments, however, are applicable quite generally.

(a) There are distinct individuals (people, representatives or whatever) gathered throughout the voting interval. The simplest supposition, in any case, is one vote, given to each (the variants such as a chairman's additional vote, tie breaking votes and so on are well known and do not alter the main point).

(b) During the voting interval, these individuals act independently. During that interval the voters do not interact either with each other or any other system. Or course, this does not prohibit all manner of debate, evidence collection, influence by the media, before the voting starts. In fact, the voting is likely to be pointless if it is not founded upon opinions shaped in some such manner.

(c) The majority logic employed, whatever it may be, is underpinned by an Aristotelian truth functional. Usually also a classical form of arithmetic is used in determining the consensual value and the calculation if performed impartially. By 'Aristotelian' logic, I mean a logic with a lattice of truth value which is a line. In a 'strict' usage, only the endpoints 'true' and 'false' are of interest but it is quite possible to have lattices with many points on the line (probabilistic logics) or midpoints (the value of 'undecided' or ' not applicable' in such a system is, however, that of a midpoint). In particular, there are no different kinds of truth values, like 'possibility' in a simple Lewis logic or the 'deontic' values, in contrast to 'alethic' values in Von Wright's 'Deontic' logic. This restriction is especially important because it excludes values such as 'self-reference' or 'existence' or 'other' or, in general, of the Boolean analogue of 'complex numbers' (see Spencer Brown, Greenberg, Hellerstein, Kaufmann, Maturana, Varela, amongst many others) needed to represent organizational closure, P-Individuation, and the like and the agreement values of coherence (for which full Günther logics seem necessary).

I do not doubt the probity of consensus and willingly conceed that it is often the best that can be done. As a rule, it is a practical and fair type of judgment and is preferable to many other types. In the case of universal accord it corresponds, uniquely, to an ideal, and there is no quibble whatsoever. But full accord is seldom encountered.

Accord, in the literal sense, is the exception rather than the rule. That being so it is useful to illustrate the potential objections to consensus, which are highlighted by accord. For this purpose there are two convenient exemplars of how it is that consensual systems differ from correspondence truth in some, at least, of all possible domains.

5. *Potential differences between coherence and consensual systems*

The first of the two exemplars is democracy, as it is commonly known in the modern and highly populated world, developed or not. Whatever the political complexion, the principles are much the same and power is ascribed (more or less wisely) to number. In contrast, there is the democracy of an ancient city state or a present day Eskimo gathering as a parliament. Probably, the parliaments of the Isle of Man and the Scandinavian countries (which differ by a matter of only one year over their date of origin, as yet undecided) have a similar ambience (or had so).

The other exemplar, to be submitted for comparison and for contrast, is science.

We consider democracy first, comparing its forms as cases in point.

In modern democracy, the count, the consensual majority, is all important and (in the worst cases) only criteria based upon (a), (b), and (c) of Section 3 are taken very seriously. People worry, for instance, over the privacy of voting booths, they show anxiety about the verisimilitude of the vote counting process and its utter impartiality. It is all very laudable, particularly when concern regarding the finite interval voting and the biasing of voters minds due to a fairly obvious media influence is taken into account. Would it not, for example, be desirable to have a voting instant ? The trouble (if any) is focused upon more primary matters, such as whether or not votes, if votes are made (which I do agree may be the best we can do) should really be independent and expedient. By way of contrast, the older states, democracies, and parliaments, expose, in their proceeding, debates that have no real limit and coherence truths of agreement (which may include agreement to disagree but to know why), where the participants are not in the least independent. Amongst other things such agreements to disagree allow the debate, often a dialectic, often far from harmonious, to continue long after gathering of participants is dissolved.

Next, consider science, let us say as 'the establishment' (or, in words of Imre Lakatos, a programme of scientific research'). The sense in which numerical democracy can be destructive to creativity is more blatantly displayed. The vote counting becomes the approval of journal editors or research councils. It may serve to stabilize the programme of scientific research (which, in mustering effort towards a just aim has every right to exist). On the other hand it obviously suppresses any form of innovation and encourages the bad habit of testing, by whatever means, those hypotheses which are minimally deviant, relatively in line with the conventional wisdom, harmless, and, on the whole, completely pedestrian, idiotic or plain malformed. This common enough experience I wish to compare and contrast with innovative, often dialectic, debate at some small seminar (a peer group, in contrast to the peer referees who seldom read but frequently reject either papers worth publication or proposals worth funding). The coherence and dispute within such groups may well result in creative leaps, genuine inventions. The phrase 'think tank' has been coined to embrace such gatherings. Unfortunately, its use indicates a desire to defuse the patent threat that the real thing brings to bear upon the established order, equally, though, the hope it offers to mankind. Maybe 'think tank' is a danger signal, maybe not.

Canons of accountability and cost effectiveness call for reasoned and well scheduled (even if delayed) response. One way of trying to obtain this result in science is to manage it by committees and their consensus (this, I submit, seldom works in science, though it possibly does in the context of an R and D project). Atkin and others have shown, quite rigorously, that most committee business is noise, meaning the circulation of irrelevancies, which, if they do anything, merely boost the status of committee members and their institutions.

There is something spurious about the consensual instant as well. Why should there be a week, day, millisecond at which something important occurs. Atkin and others have questioned the sense of asking, for example, 'when you last fell in love' or 'when Rome became a city'. With due respect to Romulus and Remus, such enquiries become even sillier as milliseconds go to microseconds go to nanosecond. The trouble is that instants of consensus do not become sillier as their situation is refined. Moments of coherence generally do. One suspects that the important moments in science should have the coherence quality.

6. The peculiar value of coherence truth and its dynamic forms

Coherence truth is about the knitting together of (usually rational) hypotheses. Rescher talked, in *The coherence theory of truth* in terms of propositions and assertoric statements. Later, in *Cognitive idealism* Rescher considers one dynamic extension which is applicable to processes generating propositions and descriptions. This is near to, if not identical with, Taylor's 'hermeneutic truth', the 'iterative refinement of meaning'. Taylor's notion is essentially kinetic and systematic, the iterative process going on amongst one or several participants, that which is refined being a statement, story, thing or event.

In conversation theory a very similar idea of dynamic coherence truth was developed in order to represent varieties of agreement and agreement to disagree between conversing participants. Conversation means concept exchange (the partial sharing of personal concepts with an agreed and shared concept in the public domain). Participants are defined in a recursive fashion, as coherent and partly autonomous systems (organizationally closed and informationally open systems), made up of stable (that is, produced from concepts and reproducible so, by means of operators that are like concepts). It also turns out, as it were in the other direction, that P-Individuated participants are conversations. They define their own boundaries, like organizationally closed autopoietic systems and the nomenclature used by Maturana, Varela, Von Foerster and my own group differs little, (for example, Maturana prefers 'structurally open and organizationally closed', when speaking of partially autonomous and interacting organisms).

Organizational Closure or P-Individuation has, itself, an hermeneutic or dynamically coherent truth value (interpretable as the existence of a stable concept, a participant, or whatever). In addition to the truth values needed to indicate the existence of (various kinds of) coherence, it is also necessary to have values of addressor and of addressee, of self-reference, of other reference. Hence, the coherence truth of an agreement does not have values that lie on a line like lattice.

7. *Differences*

So far as only formal exemplification is concerned, quite a lot can be done by a diamond shaped lattice of truth values. But for serious and practical purposes we need a more complex lattice structure accommodating, at least, the truth values of the Günther logics.

That is one respect in which consensual truth, which, as previously noted, relies upon a line like lattice, differs from coherence truth. They correspond, uniquely, only in the case of complete accord which is seldom if ever achieved. How do you know that my personal concept of something or other is your personal concept of something or other. All we do know is the shared concept in the public domain.

This matter is further complicated when we consider agreements of the type shown in the I.P.M. (of Laing) or the multiple bind situations (of Bateson). Here, we must accommodate all manner of agreements; my view of your view of something, my view of your view of my view of something and so on. All of them are perfectly legitimate, handable and realistic.

Another difference between consensus and coherence rests upon the definition of participants in a consensual or coherent domain. In the case of consensus (at least as usually considered) they are a priori distinguished individuals. So far as coherence is concerned the participants are self bounded P-Individuals, organizationally closed but informationally open. The individuals, in either case, may be people or societies or factions or representatives or institutions. No one is likely to be at odds with the fact that there are indefinitely many ways of demarcating individuals; biological, immunological, or psychological, as in P-Individuation. The method, however, differs between the cases. For consensual systems it is an observer, or observational tradition, who does the distinguishing. For a coherence system, the distinction is the boundary generated by the closure of a subsystem.

Finally, as noted before, consensual events and the events of coherence differ in their structure. With consensus, it is reasonable to talk of a consensual instant, of voting, say. Coherence does not usually occupy an instant, or any tagged interval at all.

8. Conversations of mainly directional type

Although the form is not unique, much of social and scientific dialogue is of a dialectical type. There are participants. An hypothesis known as the 'thesis' is proposed and agreed to be worth discussion. One or more opposites (not usually negations), the 'antithesis' are agreed upon by participants.

In the course of the debate, the conversation, there is 'agreement' in exactly this very liberal sense, the outcome of dialectic may be an agreement to reject one or more of the theses or the antitheses or, more commonly and productively, to create a novel 'synthesis', which may, in turn, serve as a thesis.

One caveat, over and above agreement, is that thesis, antithesis and synthesis alike, must be demonstrable. I think that demonstration means that the hypothesis, of whichever type, must have correspondence truth value in at least one possible domain, world, or universe of discourse. This domain may be physics with all of its complex and mathematically elegant substructure or the world of folk stories with all the elegant rationalization of mythologies. Or, of course, it may be a rather tedious, impoverished domain.

The dialectical conversation is a case where, a fortiori, the truth value of coherence is wedded firmly to the truth value of correspondence. But correspondence truth is to do with demonstration in some one or more possible domains, seldom if ever, in all.

9. Democracy and science revisited

The conduct of an Eskimo Parliament is one (of many) paradigms for a genuinely democratic system. It closely resembles the conduct of a conversation, often of a dialectical debate. It is frequently concerned with praxis and fact, mostly local fact, just on occasion general matters of principle. It can be shown that one way of looking at productive conversation is in terms first stated by Von Foerster, as a self-organizing system.

This, it seems, is also a primary creative mechanism in science or mathematics (inventing theorems, rather than proving them), or art. We have seen that it differs, in several ways, from a mechanism based upon consensual truth, supported by a measure of correspondence truth, or not.

The confusion between coherence based systems and consensually based systems is liable to degrade the democratic ideal. It is not that consensus is illegitimate, or badly twisted (on the whole, it is not, though it may be). The trouble is that if consensus and the presuppositions which underlie it are mistaken for coherence, there is grave danger of undue fixity (clearly, there must be some ossification, some stability), and regression to the norm. The rape of democracy, which I have complained about and ascribed to this confusion, is really the abdication of democracy, to mediocrity. Or, for that matter, the abdication of science to mainstream technology.

10. Compatibility of the mechanisms

The hazard does not lie in consensus. Often consensus is the best that can be done. It is often taken for granted that real democracies become impossible when the population is large or the questions at issue are urgent. Consensus is a good means of lightening the administrative burden and offering a somehow fair form of government.

The real hazard does lie in mistaking one mechanism for the other. If it is the best we can do, then there is good reason to invoke a consensual process (like voting, in certain years on a given day) provided that this is recognized as consensual type resolution and provided also that there is encouragement for real debate, coherence evaluation, and so on in between (which is just what does not occur, if consensus is mistaken for coherence and democracy).

11. Is it only the best

In my view, we can probably do better than consensual resolution, interspersed between coherence valued discussion. This view is realistic because of the increasing dominance of information, knowledge dissemination, computation, communication and the like in society. Relatively, all of these capabilities (hitherto prohibitively costly and unreliable) are inexpensive. Using these resources to promote conversation (rather than mere data passing), it should be possible to maintain larger scale agreements or coherencies involving more and more distributed societies.

Forsythe, in education; Robinson in managerial organization; we in Amsterdam, are actively pursuing this possibility.

12. Beyond technology

Up to date technology is only one prerequisite for sustaining what Robinson calls conversations between societies. One other at the least is required.

This one is a way of seeing, catalyzing, and manipulating the knowing which occurs in conversation and of representing those shared concepts, in the public domain, which may count as being knowledge. All that hinges upon some theoretical work (which exists) but, of even greater importance, an increased respect for creating those beautiful analogies, allegories, stories and pictures (some rich in mathematical symmetries, some in poetry) that are designated, as agreements are, by metaphors.

* This article was first published in *Communication & Cognition, vol. 23, nos. 2-3.*

References

Bateson, G., 1973, *Steps towards an Ecology of Mind*, Paladin, New York.

Günther, G., 1962, Cybernetic Ontology and Transjunctional Operations, in *Self-organizing systems*, M. Yovits, G. Jacobi and G. Goldstein (eds.), Spartan Books, Washington, pp. 313-393.

Hellerstein, N.S., 1985, Diamond, a logic of paradox, *Cybernetics*, vol. 1, no. 1, pp. 101-115.

Kauffman, L., 1986, Self-reference and Recursive Forms, *Journal of Social-Biological Structures*, no. 10, pp. 53-67.

Lakatos, I., 1968, Criticism and the Methodology of Scientific Research Programmes, *Proc. of the Aristotelean Society*, vol. 69, pp. 149-186.

Laing, R., 1961, *Self and Others*, Tavistock, London.

Maturana, H.R., 1975, The organization of the living: a theory of the living organization, *International Journal Man-Machine Studies*, vol. 7, pp. 313-332.

Maturana, H.R.; Varela, F.J., 1980, *Autopoiesis and Cognition*, Reidel, Boston.

Pask, G., 1975, *Conversation, Cognition and Learning*, Elsevier, Amsterdam.

Pask, G., 1975, *Conversation Theory, Applications in Epistemology and Education*, Elsevier, Amsterdam.

Pask, G., 1980, Developments in Conversation Theory, *International Journal Man-Machine Studies*, vol. 13, pp. 357-411.

Pask, G., 1980a, The Limits of Togetherness, in *Information Processing 80*, S. Lavington (ed.), North Holland, New York, pp. 999-1012.

Rescher, N., 1973, *The Coherence Theory of Truth*, Oxford University Press, London.

Rescher, N., 1983, *Cognitive Idealism*, Oxford University Press, London.

Spencer Brown, G., 1969, *The Laws of Form*, Allen and Unwin, London.

Taylor, C., 1964, *The explanation of behaviour*, Routledge and Kegan Paul, London

Taylor, C., 1973, Interpretation and the sciences of Man, *Review of Metaphysics*, vol. 25, pp. 3-51.

Varela, F.J., 1975, A Calculus for Self-reference, *International Journal for Genereal Systems*, vol. 2, pp. 5-24.

Varela, F.J., 1979, *Principles of Biological Autonomy*, North Holland-Elsevier, Amsterdam.

Von Foerster, H., 1981, *Observing Systems*, InterSystems Publications, The Systems Inquiry Series, California.

Von Wright, G.H., 1963, *Norm and Action*, The MIT Press, Cambridge, Massachusetts.

WRITERS OF THE LOST I: SECOND-ORDER SELF-OBSERVATION AND ABSOLUTE WRITERSHIP

Eric Rosseel

Satie
Free University of Brussels (VUB)

Abstract

The present paper sketches a framework in order to understand a pattern of absolute writership that is centred around the notion of 'second-order self-observation', i.e. self-observation that looks at its own instruments of self-observation. Particularly, the life and works of the Portugese author Fernando Pessoa have been used as a starting point. The habit of 'second-order self-observation' has been analyzed as an artificial deviation from the natural emergence of consciousness in the breakdown of some action process, and to the degree that the actor (or non-actor) reports to suffer and longs for self-destruction, the habit can be labeled 'pathological'. The analysis ends up in an a more detailed consideration of the various aspects of absolute writership.

1. Introduction

This paper has been inspired by our aim to understand the writership of the Portugese futurist Fernando Pessoa (1888-1935). When reading his work and getting acquainted with his biography, we were struck by some aspects that happened to characterize a lot of other writers (such as Baudelaire, Cesare Pavese, J-K. Huysmans, and, more particularly, Pessoa's friend and companion Mario de Sá-Carneiro). First of all, we were of course confronted with his heteronymy, the fact that he multiplied himself in some fifty different authors, the most famous being Alberto Caeiro, Ricardo Reis, Alvaro de Campos, Bernardo Soares and Fernando Pessoa himself. This multiplication of personalities was an attempt to grasp the mystery that surrounded the opposition between the observed world (as Reality and Fiction) and himself as observer and spectator. The permanent obsession with the mystery of reality and the impotence of his Intelligence to understand the true nature of things (the theme of Pessoa's Faust), provoked a

G. Van de Vijver (ed.), New Perspectives on Cybernetics, 233–245.

longing for Sleep and Death, a desire to lose Consciousness. In this sense, his self-observation is '*self-destructive*'. This Baudelairian *Spleen* is described at length in '*The Book of Intranquillity*' (Pessoa, 1982 - under the heteronym of Bernardo Soares). This 'self-destructiveness' sealed his fate: in 1935 he died from liver cirrhosis due to excessive drinking.

Secondly, Pessoa was an absolute writer: his whole life was in fact a *literary* life. In no way a distinction can be made between his life and his work. His love relationship with 'Ophelia' was complicated by the 'interventions' of his heteronym Alvaro de Campos and, finally, he broke off his relation with Ophelia because he had to devote himself to 'some higher vocation'. Pessoa had no real profession, working a few hours a week for a commercial firm in order to gain some income. Moreover, he had no family responsibilities. As was the case for the Italian suicidal writer Cesare Pavese, he felt himself an outsider, unable to live a 'normal' civilian life; whereas Pavese suffered psychologically from this feeling of social incompetence, Pessoa suffered metaphysically. And thirdly, we were really astonished by the fact that nearly all the authors (not only Pessoa but also a lot of other writers such as Pavese, Sá-Carneiro, Poe, Baudelaire, Rimbaud, Maiakowski, Sylvia Plath, Rilke, Huysmans, ...) whom we recognized as 'self-destructive self-observers', lost their father very early in their life (in Rimbaud's and Rilke's case, the father did not die but was physically absent).

Through these three elements ('self-destructive' self-observation, absolute writership and orphanage), we felt able to understand Fernando Pessoa in his particularity and singleness, even if we could hardly formulate this intuitive conceptual framework. His preoccupation with the basics of personality (as for instance in his Alvaro de Campos' text Ultimatum) and with his unslaked desire to be conscious of his loss of consciousness (as in sleeping or day-dreaming), helped us to solve the problem of Selfhood and Person(Ego)hood by stressing the discontinuous linguistic character of the Ego: the Ego is a self-observing speech act, a momentum of speaking. And as such the self-observing speech act may stand on its own or form part of an integrative discourse (Rosseel, 1990).

The linkage between self-observation and absolute writership lies in what we have called the *Ceifeiric Delusion* (Rosseel, 1989). 'Ceifeira' is one of Pessoa's verses most preferred by him, in which he expresses his envy of the joyful 'unconscious' activity of a harvest woman (Pessoa, 1960):

Ah! If I could be you, at the same time being myself
Have your gay unconsciousness
And consciousness as well!

These verses express a paradoxical desire to lose oneself and at the same time to *know* that one has lost self-consciousness. In many other verses and more particularly in his 'subjective tragedy' Faust, Pessoa words this desire towards ignorance and innocence, the desire to escape from the painful consciousness of the unattainability of that desire. So, the paradox reveals us that Pessoa was a self-observer in nearly all moments of his life, and that he *suffered* from his excessive self-observation.

So, the paradox refers to a wider question: *How can self-observation, i.e. talking about one's self, turn into a 'pathology' of 'self-destructive self-observation' ?* We therefore shall develop the notion of '*second-order self-observation*' as a faculty of linguistic self-observing systems.

2. *Action and Self-Observation*

A self-observing system is a system that is capable of engaging into observations in which the distinction is made between the observing system (called the Self) and other systems (Others, Objects or the World in general). By making the distinction between Self and 'Other', the system handles some 'image' of itself. A self-observing system thus is capable of speaking about oneself, of 'knowing oneself'. A self-observing system 'has' a *discursive Self*. What do we precisely mean by 'second-order self-observation', and how can a self-observing system develop 'second-order self-observation' and evolve into a system that eventually destroys itself (its Self) ? To answer this question, we propose to turn our attention to the way *first-order* self-observation originates.

Our 'lecture' of Pessoa then is framed in a radical pragmatist-constructivistic conception of (self)-consciousness. Our analysis is inspired by the famous James/Mead distinction between 'I' and 'Me' (James, 1890; Mead, 1990), the thinking of the Chilean biologist Humberto Maturana (Maturana & Varela, 1980) and, finally, Julian Jaynes' underestimated historical account of the breakthrough of personal consciousness (Jaynes, 1976). All of them reject the basic dualism of mind-body or subject-object that since Descartes continues to mystify the nature of consciousness, and they look at the emergence of consciousness in the context of the action of a human organism as a system of structural plasticity that is coupled to an ambience. Then, the theme of self-consciousness refers to three important questions:

1. the specification of a world of objects (including one's Self) that are perceived as unities existing independently of ourselves as actors and observers.

2. the origin of consciousness as a means to overcome the inhibition of action in circumstances wherein the actual structural potentialities of a human system are inadequate.

3. the eventuality of a 'pathological' evolution of self-awareness into a form of depressive self-observation that is detached from action in and upon the world.

We start from the viewpoint that action is inherent to all living organisms and that it does not depend upon the development of sensory, neuronal or motor organs. So, cognition can be conceived in terms of *competence to action* rather than in terms of instructive mental representations of an outer reality (Maturana, 1988). Consciousness (or more general 'perception') thus does not precede action, it emerges when action is blocked by some unanticipated event, a so-called *breakdown* (Winograd & Flores, 1986; Bersini, 1990). In these moments of breakdown, consciousness arose and arises as the production of imagery and language, i.e. of 'objective' distinctions. The emergence of coordinated imagery and language may be situated at two levels: first, at the level of systems in which language and imagery are already firmly established; secondly, the development of self-consciousness in ontogeny and phylogeny.

The first level presents few difficulties. Self-observation occurs when the structural coupling of a human system to a specified ambience gets blocked by an environmental perturbation that is compensated by a 'breakdown' of the action of the system. As the system does not dispose of an adequate action that would have continued his drift in that ambience, 'mindless' automated action is interrupted and self-observation helps to explore the plasticity of the system and to develop a new 'creative' action that removes the blocking (Norman & Shallice, 1986; Von Cranach et al., 1986). Only in those moments, we do see what we see, we look at the world and talk about the images we have formed; only in these moments, objects arise as realities created in language and imagery. Objects emerge as *pretensions*, as something we pre-tend, i.e. stretch out in front of us, something we hold in our hand and bring to our eyes. In moments of 'stress', we look to the 'object' (or the 'situation') as something that we do not know and cannot name: we have no action to perform upon it. By looking at it 'as it is in itself', we analyze it, relate it to other objects and so give it meaning. In other words, we act with Intelligence.

The question of the emergence of imagery and language in ontogeny and phylogeny cannot be answered precisely. However, we know that the genesis of Man is linked to a radical breakdown in the food gathering techniques of the apes when climate changes shrinked the tropical forests (beginning from about 15 million years ago) and forced some hominids to survive in the open planes of the savannahs. These changes allowed a transition from kinesthetic motor feedback to *visual* feedback and enabled primitive man to renounce the immediate food value of things and to look at objects as potential instruments for food gathering. The new life situation of man fostered an *'objective' visualization* of the perturbations that triggered him. Human activities were social activities (apes do not share food) and so they stimulated the use of language that evolved from an animal sign language to a language that served to coordinate new activities that had to be transmitted from generation to generation. The dominance of the eye-hand coordination makes the world into a world that can be looked at and that can be named, about which can be spoken. Remembrance, as the seeing and the hearing of absent things and persons in a kind of perceptions that resemble hallucinations, begins to develop. As Julian Jaynes has so convincingly documented, in these visions a person receives instructions how to handle things; after a long process of cultural changes, Man finally began to see himself as the *agent of his Self* rather than to experience himself as a machine inspired by good and evil gods (Jaynes, 1976).

In the individual ontogeny of the child, an analogous process of personalization results from the socialization efforts of the parental environment. As a child, we are introduced in the world of images and words. In the first important phase of socialization, our parents and relatives point to the world and learn us to produce images and to make distinctions. It has been argued that, indeed, perceptions comprise a propositional (thus linguistic) and a qualitative (imaginative) content (Natsoulas, 1974; Neisser & Kerr, 1973). 'This is Daddy!' is an invitation to *look* at Daddy, to distinguish him as a figure against a background, and, at the same time, to *name* him. Through imagery and language, the child succeeds in specifying a world of objects that it shares with its social environment but that also very quickly is personalized. Thus (at least this is what we have tried to argue here), as well in anthropogenesis as in the ontogenesis of the human, individual imagery and language play a decisive role. In both cases, consciousness and self-consciousness appear in the context of breakdowns, situations in which we have no ready-made action available and in which our coupling to the ambience is questioned (by the parents in the case of the infant). We have learned, through our ontogeny and phylogeny, to specify in such moments a vision of things as objects in front of us, objects that stand against us; we cease to be *in* the world and start to look *at* the world, creating objects as 'order from chaos'. And as we see the objects, we see ourselves as the observer of these objects: self-awareness is born.

3. *Self-destructive Self-Observation*

In the foregoing analysis, the Ego emerged in living systems as a realization of an imaginary-linguistic potentiality of an observing system so that its viability in a changing ambience is guaranteed. Self-observation leads to the development of new actions manifested in self-steering. Creativity can be obtained in two ways. First of all, simple *innovation*: the expression or invention of something 'new' by recombining the elements of a given language. An innovation will be recognized very easily because we master the constituting elements of the code upon which the new product is based. So, an innovation will be at the same time recognized as new and familiar, it is not at all strange. Another form of creativity is reached when a person does not simply recombine elements but plays with the constituting elements of language itself, e.g. by altering the 'definitions' of these elements. Now, public and audience are confronted with the arbitrariness of linguistic codes; the absence of an ontological relation between an 'object' (this may also be a gesture) and its meaning is uncovered. The object is experienced as *strange*. This is what happened in futurism, dadaism, abstract painting, minimal music, conceptual art and so on. In our opinion, simple innovation is linked to 'first-order' self-observation, whereas the second form ('art') results from 'observing *too long*', from what we have called '*second*-order self-observation'.

The expression 'observing too long' suggests that at least in some cases self-observation may deviate from its 'natural' course, and that such a deviation is 'pathological' or at least 'artificial'. We had liked to emphasize that in our use the word 'pathological' refers to a purely *subjective* evaluation in that the person himself reports to suffer, it has no real medical or moral connotation. We simply mean that self-observation may be detached from the problem of disinhibition of action and may become an act on its own. With regard to the biological viability of the system, such 'artificial' self-observation may be either self-*instructive* or self-*destructive*. Artificial self-observation may be explored for its own sake, simply in order to enjoy the contemplation of things and the experience of beauty. In this form of self-instructive self-observation, no context of breakdown of action is present. Self-*des*tructive self-observation, however, originates in a failure to overcome the breakdown satisfactorily.

Indeed, in the process of a breakdown of the structural drift of a (human) system, visual imaginary activity looks in our memory for a solution for the blockade of our actions. If such a 'creative' or 'reproduced' solution is found, the stream of action is picked up again, the person gets absorbed in the drift of his occupations, self-awareness is no longer devastated by the subject-object distinction and the 'I' fades away (however 'perception' continues). This is what we mean by the Discontinuity of the Ego in the Continuity of the Self.

However, in some cases the person is not able to initiate new actions and he gets stuck fast in the observation of the elements of his problem (including his idle hands). Instead of being a source of new intelligent interactions with the world, self-observation becomes a problem that troubles the system and undermines its viability. The world of objects may be experienced as threatening and self-observation may be used as a defence against feelings of guilt. In the long run, syndromes of depression and alienation become dominant and the person complains of boredom and inactivity, thus from more or less permanent self-observation.

The basic condition of the pathology of self-destructive self-observation consists in the fact that the link between self-observation and the recovery of action is broken. This inhibition of action results from an internal conflict between acting and non-acting. A kind of 'frozenness' of the senses sets in: one does no longer look at the things of the world, but stares and gazes at them. The effort put in searching to find a new action in front of the perturbation provoked by the ambience, may end up in *anxiety*, *excitement* or otherwise in *apathic contemplation*, when such effort is given up. These end results may become a habitual mode of uncoupling from one's social ambience. Anxiety leads to a collapse of the finer distinctions within the Non-Self, prevents the emergence of any adequate action and, so, provokes the sensation of Death and Nothingness. In contemplation, the analysis of the concrete problem of action is transformed into a contemplation of the objects of the world inclusive of oneself that contemplates these objects. Thus, a *second-order self-observation* is performed upon the distinction between Self and World that arose in the first-order self-observation. Self-observation no longer aims at restoring the interaction with the world but gets lost in a 'philosophical' investigation of the 'true' metaphysical nature of Subject and Object.

The above analysis can be brought in line with the work of the English child psychoanalyst Donald Winnicott (1964, 1985) and the German therapist Alice Miller (1979, 1981). Winnicott observes among children that show a problematic insertion in the social world a kind of *precocious intelligence*. Some of these children get submerged by external stimulations, and they only dispose of their mental faculties to master their high levels of excitement. This results in a precocious hypermaturation of the mental faculties, including 'self-observation'. Winnicott's work illustrates that mental activity can become an 'object to itself' that no longer regulates the interaction with the world. Children that grow up under such conditions may develop excessive daydreaming, lose themselves in reading books and looking at pictures and so may become very talented artists and thinkers (as the French say: 'Ça fera de la bonne littérature!'). On the other hand, Alice Miller holds that most mothers and parents in general use their young children to satisfy their own needs for warmth, recognition and self-esteem. So, the child learns, by fear of loss of love, to deny and repress (i.e. fails to 'speak' and 'dialogue with') his of her feelings of anger, anxiety and sadness. His or her intelligence is oriented towards the development of a False Self that protects him against his unbearable feelings of love and hate. The conflict between satisfying the parental desire (the False Self) and the unworded desires of the Real Self provoke vague feelings of unrest and dissatisfaction which are not understood and which develop into a penetrating sentiment of *Ennui*.

4. Writing as self-instruction

The 'pathetic self-observer' continuously confronts his problematic insertion in the world of things. In his habit of non-action he feels the gap that his self-observation has driven between his eyes and the things he is looking at. While action realizes the structural plasticity of a system through the coupling to new ambiences and thus ends up in self-construction, the pathetic self-observer must turn to other solutions to overcome the gap that separates him from being-in-the-world. As he is a master in making distinctions, in mastering the world by *naming* things, it is quite evident that he turns to literature and art to create himself a universe in which he comes to master the monsters that language has bred. In other words, in his search for his Real Self (Arthur Rimbaud: 'JE est un Autre!' - I is an Other!), he *instructs* himself to be someone else. Edgar Allan Poe's search for the bizarre ended up un Baudelaire's dandyism, the search for the most artificial, refined, differentiated and individualized sensations, in the decadence of Oscar Wilde's Dorian Gray and J-K. Huysmans' Des Esseintes (who, by the way, served as a model for Dorian Gray). All these writers try, in their writings, to find 'something else', to sojourn in a land 'somewhere else', as Pessoa's friend Sá-Carneiro expressed this lamentation. And Baudelaire stated firmly: 'I think I will always be happy there where I am not!'. On the other hand, there is the belief in an *artificial* Self, as a source of genuine Beauty: in his famous poem 'Autopsychography' Pessoa characterizes the poet as someone who *feigns*, and adds: 'To feign is to know oneself!'.

The *doubling* that is implied in self-observation is well illustrated by the following of Pessoa's poems:

The child played/ With a little cart
He felt himself playing/ And he exclaimed: I am two!

There is one who plays/ And another who knows it.
One sees me playing/ And the other sees me looking

I am there behind myself;/ But when I turn my head
I was not what I believed to be/ To turn around is just that

The other boy/ Has no legs nor arms
It is not a child/ He has no mother nor brothers

(...)

And the other who lives in me/ And who feels with me,
Is not my father, nor my uncle,/ Not my body, not my friend

He has a soul here at home/ Sees me without looking at me,
And the little cart/ Becomes apparent.

This poem reveals the troubling consequences of self-observation. Reflection does not integrate the playing; from the very first verses on, the playing is observed from a distance. And consequently, action is interrupted in the third strophe. From there on, reality is split: the child that was playing versus the child that sees and knows; the child versus the little cart. The uncanny consequences of the doubling involved in second-order self-observation may thus be resumed in the confusing question 'Who am I, that I that is able to ask this question ?' and in the consciousness of the *exteriority* of things ('and the little cart becomes apparent'). The suffering begins when the subject observes that his search for the true nature of the distinctions that he makes, only augments his incapacity to engage in the world of things. But in his intelligence, the subject (Pessoa) is unwilling to accept the impossibility to understand the mysteries of life. So, in his persistence of observing things as exteriorities from which one is separated, the observer becomes a *spectator* who puts himself aside of life. He damns himself to be no-one. In the permanent self-dialogue that he holds up, the idea of the Ego as an other, as someone else, may emerge. In his frustrated mood, the subject tries to reach a new way of being or to remember an original way of being that went lost. When this search for different modes of sensing the world is strived for by a *multiplication of the personality* (as Baudelaire advocated with respect to the use of wine and hashish) so that new personalities are filled up with fictional elements, heteronyms may be born.

5. Deregulated life and absolute writership

Absolute writership is characterized by some particular form of unhappiness, some way of disagreeing with one self. There is striking incapacity (or unwillingness) to act and to insert one's daily activities into the stabilized patterns of societal life, even if some of the poets that we have in mind, may have periods of revolutionary eruptions (e.g. Baudelaire in 1848) or even a personality of revolutionary excitement (e.g. Maiakowski). Pessoa, Pavese, Baudelaire, Rilke, all of them had no real professional life, they were 'politically' inactive or naive, they did not found a family, they had difficulties in making contacts with others and so on. This incapacity to act is linked to a form of semi-permanent observation of the world of things and of oneself as the observer of these things. The relation with the world of things always remains problematic, a source of dissatisfaction because its true nature cannot be grasped, in other words, cannot be transformed into language.

Self-observation as linked to the capacity to 'speak' about the true nature of things and of oneself as the observer of these things, is a source of intense suffering. In the self-observation of that what is most human, namely (first-order) self-consciousness, *man cannot stand himself.* Two ways are open to alleviate the pain of existence. One solution is to *eliminate consciousness* by drugs or other addicting activities. The other solution is the one J-K. Huysmans has chosen after writing his most 'decadent' *A Rebours* (the book that inspired Oscar Wilde to write his Dorian Gray). It is the solution of *mysticism*: unable to integrate collectivities, the self-observer tries to feel himself part of the total Universe, of the All and the Absolute, the kick of looking God into his eyes, of feeling the timeless Divine or Diabolic inside one's Self.

Summarizing, we could hold that the absolute writer, in order to escape from the 'spleen' of everyday life and the banality of society, tries to take very seriously Rimbaud's advice to *deregulate all the senses* and to become a *seer*. But he loses his way in this dramatic attempt to see the mystery instead of solving it, as Heidegger would say.

So, the conception of absolute writership in terms of second-order self-observation synthesizes a whole pattern of 'metaphysical' attitudes and biographical dynamics. As a conclusion we had liked to advance this *pattern of absolute writership* in twelve points.

1. *A dramatic relationship with the parents.* In most cases, an affective disillusion vis-à-vis the mother and an absence of the father from late childhood on. Many of Pessoa's verses that treat about childhood, associate it immediately

with orphanage. Pessoa's psychical instability is mostly associated with the sudden death of his father in 1893, followed six months later by the death of his one year old brother Jorge. These tragic events may not make us forget that his mother was rather cool, proud and emotionally frugal. In an Autobiographical Note written in 1907 he writes that he cannot entrust his problems and his need to complain himself to his mother. The development of self-dialogue may well be intensified when significant persons in a child's life, particularly the father, 'disappear'. The void that is left by the father, may frustrate the child's need to dialogue with an Other, and so his wish to be himself that Other that is absent. Hypermature intelligence may undertake an intense work of imagination that takes on the form of heteronymy: the child puts himself in the position of an imagined other, eventually objectified in literary creation. For a 'romantic' description of this first point, one is recommended to read the chapters on the 'youth of the poet' in Milan Kundera's 'Life is Elsewhere', the history of a fictitious poet based on the lives of Rimbaud, Baudelaire, Rilke, Maiakowski and maybe many others.

2. *A precocious intelligence.* Mostly excellent school results, vivid imagination and refined language-sense.

3. *Awareness of the existence of things and the absurdity of being.* From his early youth the absolute writer questions the nature of things and develops the habit of daydreaming.

4. *Intensive self-observation.* Cf. 3.

5. *Feelings of separateness from 'true' reality.* This is a direct consequence of points 3 and 4.

6. *Refusal to insert oneself societally.* The absolute writer is rather a-social, often clumsy in his social contacts, dislikes immediate physical contact. He wants to be different, sometimes feels himself superior. At the other hand, he does not fit the 'normal' pattern of civilian life, as already has been argued at length.

7. *Impotence to lose oneself in an erotic relation.* The absolute writer has an ambivalent stance against erotism. He idealizes Feminity (or Masculinity) but is disappointed when his Idea is not materialized in the contact with a real, concrete woman. He desires to feel unified in harmony, but is (due to points 4 and 5) unable to realize this union. Problems of impotency may result (Pavese).

8. *Sexual inhibitions and insecurity.* Cf. 7.

9. *Desire for self-negation.* The writer tries to adopt a different personality. He takes on a pose (Rilke), feigns (Pessoa), chooses to be a dandy (Baudelaire) or a decadent (Huysmans). Or tries to find a compensation for the lack of social bonds in mysticism (Pessoa, Huysmans).

10. *Writing in substitution for an active life.* Writing is not meant as the proliferation of a message, an attempt to convince others or to share feelings of beauty, solidarity; it is not intended as *communication*, that aims to have an effect upon the reader. Writing is, as the dadaist Tristan Tzara would say, a 'poetical activity of the spirit'.

11. *Death: the ultimate solution for the problem of knowledge.* The endless search to lose oneself, to be unified with the resisting things, to feel their true nature finally provokes deep despair. The ultimate solution of death, as the eternal reconciliation with the infinitude of things, solicits more and more attention.

12. *Physical death during midlife.* Pessoa aged 47, Pavese 42, Baudelaire 43. Huysmans was 36 when, after writing A Rebours, two solutions remained open: either to commit suicide, or to become a catholic.

Time and space do not permit us to illustrate these points more deeply and systematically. Moreover, it is not our intention to build a logical theory of types of writers. We hope that we have developed a better understanding of Pessoa and other absolute writers, a way to *cognize*. So, we can learn to live better with them, in other words, to read them more coherently.

References

Bersini, H., 1990, Before and after breakdowns, *Proceedings of the 4th Colloquium of the 'Association pour la Recherche Cognitive*, Paris, 28-30 March 1990.

James, W., 1890, *Principles of Psychology*, Henry Holt, New York.

Jaynes, J., 1976, *The Origin of Consciousness in the Breakdown of the Bicameral Mind*, Houghton-Mifflin, Boston.

Maturana, H.R., 1988, Reality: the search for objectivity or the quest for a compelling argument, *Irish Journal of Psychology*, vol. 9, pp. 25-82.

Maturana, H.R.; Varela, F.J., 1980, *Autopoiesis and Cognition: the Realization of the Living*, Reidel, Dordrecht.

Mead, G.H., 1934, *Mind, Self and Society*, University of Chicago Press, Chicago.

Miller, A., 1979, *Das Drama der begables Kind*, Suhrkamp, Frankfurt.

Miller, A., 1981, *Du sollst nicht merken*, Suhrkamp, Frankfurt.

Natsoulas, T., 1974, The subjective experiential element in perception, *Psychological Bulletin*, vol. 81, pp. 611-631.

Neisser, U.; Kerr, N., 1973, Spatial and Mnemonic Properties of Visual Images, *Cognitive Psychology*, vol. 5, pp. 138-150.

Norman, D.A.; Shaliice, T., 1986, Attention to Action: willed and automatic control of behaviour, in *Consciousness and Self-regulation*, vol. 4, R.J. Davidson et al. (eds.), Plenum, New York.

Pessoa, F., 1960, *Obra Poética*, Nova Aguilar, Rio de Janeiro.

Pessoa, F., 1982, *Livro do Desassossego* (The Book of Intranquillity), Atica, Lisboa.

Rosseel, E., 1989, *The Ceifeiric Delusion: a theory of the I as part of the structural drift of the Self*, Free University Brussels.

Rosseel, E., 1990, Notes on Self and Self-steering, in *Self-steering and Cognition in Complex Systems: Toward a new Cybernetics*, F. Heylighen, E. Rosseel and F. Demeyere (eds.), Gordon & Breach, London.

Von Cranach, M.; von Ochsenbein, G.; Tschan, F., 1986, Action of social systems, in *Perspectives on Contemporary German Social Psychology*, G.R. Semin and B. Krahé (eds.), Sage, London.

Winnicott, D.W., 1964, *The Child, the Family and the Outside World*, Penguin Books, Harmondsworth.

Winnicott, D.W., 1985, *The maturational processes and the facilitating environment: studies in the theory of emotional development*, Hogarth, London.

Winograd, T.; Flores, F., 1986, *Understanding Computers and Cognition*, Ablex, New York.

INDEX

SYNTHESE LIBRARY

Studies in Epistemology, Logic, Methodology, and Philosophy of Science

1. J. M. Bochénski, *A Precis of Mathematical Logic*. Translated from French and German by O. Bird. 1959 ISBN 90-277-0073-7
2. P. Guiraud, *Problèmes et méthodes de la statistique linguistique*. 1959 ISBN 90-277-0025-7
3. H. Freudenthal (ed.), *The Concept and the Role of the Model in Mathematics and Natural and Social Sciences*. 1961 ISBN 90-277-0017-6
4. E. W. Beth, *Formal Methods*. An Introduction to Symbolic Logic and to the Study of Effective Operations in Arithmetic and Logic. 1962 ISBN 90-277-0069-9
5. B. H. Kazemier and D. Vuysje (eds.), *Logic and Language*. Studies dedicated to Professor Rudolf Carnap on the Occasion of His 70th Birthday. 1962 ISBN 90-277-0019-2
6. M. W. Wartofsky (ed.), *Proceedings of the Boston Colloquium for the Philosophy of Science, 1961–1962*. [Boston Studies in the Philosophy of Science, Vol. I] 1963 ISBN 90-277-0021-4
7. A. A. Zinov'ev, *Philosophical Problems of Many-valued Logic*. A revised edition, edited and translated (from Russian) by G. Küng and D.D. Comey. 1963 ISBN 90-277-0091-5
8. G. Gurvitch, *The Spectrum of Social Time*. Translated from French and edited by M. Korenbaum and P. Bosserman. 1964 ISBN 90-277-0006-0
9. P. Lorenzen, *Formal Logic*. Translated from German by F.J. Crosson. 1965 ISBN 90-277-0080-X
10. R. S. Cohen and M. W. Wartofsky (eds.), *Proceedings of the Boston Colloquium for the Philosophy of Science, 1962–1964*. In Honor of Philipp Frank. [Boston Studies in the Philosophy of Science, Vol. II] 1965 ISBN 90-277-9004-0
11. E. W. Beth, *Mathematical Thought*. An Introduction to the Philosophy of Mathematics. 1965 ISBN 90-277-0070-2
12. E. W. Beth and J. Piaget, *Mathematical Epistemology and Psychology*. Translated from French by W. Mays. 1966 ISBN 90-277-0071-0
13. G. Küng, *Ontology and the Logistic Analysis of Language*. An Enquiry into the Contemporary Views on Universals. Revised ed., translated from German. 1967 ISBN 90-277-0028-1
14. R. S. Cohen and M. W. Wartofsky (eds.), *Proceedings of the Boston Colloquium for the Philosophy of Sciences, 1964–1966*. In Memory of Norwood Russell Hanson. [Boston Studies in the Philosophy of Science, Vol. III] 1967 ISBN 90-277-0013-3
15. C. D. Broad, *Induction, Probability, and Causation*. Selected Papers. 1968 ISBN 90-277-0012-5
16. G. Patzig, *Aristotle's Theory of the Syllogism*. A Logical-philosophical Study of *Book A* of the *Prior Analytics*. Translated from German by J. Barnes. 1968 ISBN 90-277-0030-3
17. N. Rescher, *Topics in Philosophical Logic*. 1968 ISBN 90-277-0084-2
18. R. S. Cohen and M. W. Wartofsky (eds.), *Proceedings of the Boston Colloquium for the Philosophy of Science, 1966–1968, Part I*. [Boston Studies in the Philosophy of Science, Vol. IV] 1969 ISBN 90-277-0014-1
19. R. S. Cohen and M. W. Wartofsky (eds.), *Proceedings of the Boston Colloquium for the Philosophy of Science, 1966–1968, Part II*. [Boston Studies in the Philosophy of Science, Vol. V] 1969 ISBN 90-277-0015-X

20. J. W. Davis, D. J. Hockney and W. K. Wilson (eds.), *Philosophical Logic*. 1969 ISBN 90-277-0075-3
21. D. Davidson and J. Hintikka (eds.), *Words and Objections*. Essays on the Work of W. V. Quine. 1969, rev. ed. 1975 ISBN 90-277-0074-5; Pb 90-277-0602-6
22. P. Suppes, *Studies in the Methodology and Foundations of Science. Selected Papers from 1951 to 1969*. 1969 ISBN 90-277-0020-6
23. J. Hintikka, *Models for Modalities*. Selected Essays. 1969 ISBN 90-277-0078-8; Pb 90-277-0598-4
24. N. Rescher *et al.* (eds.), *Essays in Honor of Carl G. Hempel*. A Tribute on the Occasion of His 65th Birthday. 1969 ISBN 90-277-0085-0
25. P. V. Tavanec (ed.), *Problems of the Logic of Scientific Knowledge*. Translated from Russian. 1970 ISBN 90-277-0087-7
26. M. Swain (ed.), *Induction, Acceptance, and Rational Belief*. 1970 ISBN 90-277-0086-9
27. R. S. Cohen and R. J. Seeger (eds.), *Ernst Mach: Physicist and Philosopher*. [Boston Studies in the Philosophy of Science, Vol. VI]. 1970 ISBN 90-277-0016-8
28. J. Hintikka and P. Suppes, *Information and Inference*. 1970 ISBN 90-277-0155-5
29. K. Lambert, *Philosophical Problems in Logic*. Some Recent Developments. 1970 ISBN 90-277-0079-6
30. R. A. Eberle, *Nominalistic Systems*. 1970 ISBN 90-277-0161-X
31. P. Weingartner and G. Zecha (eds.), *Induction, Physics, and Ethics*. 1970 ISBN 90-277-0158-X
32. E. W. Beth, *Aspects of Modern Logic*. Translated from Dutch. 1970 ISBN 90-277-0173-3
33. R. Hilpinen (ed.), *Deontic Logic*. Introductory and Systematic Readings. 1971 *See also* No. 152. ISBN Pb (1981 rev.) 90-277-1302-2
34. J.-L. Krivine, *Introduction to Axiomatic Set Theory*. Translated from French. 1971 ISBN 90-277-0169-5; Pb 90-277-0411-2
35. J. D. Sneed, *The Logical Structure of Mathematical Physics*. 2nd rev. ed., 1979 ISBN 90-277-1056-2; Pb 90-277-1059-7
36. C. R. Kordig, *The Justification of Scientific Change*. 1971 ISBN 90-277-0181-4; Pb 90-277-0475-9
37. M. Čapek, *Bergson and Modern Physics*. A Reinterpretation and Re-evaluation. [Boston Studies in the Philosophy of Science, Vol. VII] 1971 ISBN 90-277-0186-5
38. N. R. Hanson, *What I Do Not Believe, and Other Essays*. Ed. by S. Toulmin and H. Woolf. 1971 ISBN 90-277-0191-1
39. R. C. Buck and R. S. Cohen (eds.), *PSA 1970*. Proceedings of the Second Biennial Meeting of the Philosophy of Science Association, Boston, Fall 1970. In Memory of Rudolf Carnap. [Boston Studies in the Philosophy of Science, Vol. VIII] 1971 ISBN 90-277-0187-3; Pb 90-277-0309-4
40. D. Davidson and G. Harman (eds.), *Semantics of Natural Language*. 1972 ISBN 90-277-0304-3; Pb 90-277-0310-8
41. Y. Bar-Hillel (ed.), *Pragmatics of Natural Languages*. 1971 ISBN 90-277-0194-6; Pb 90-277-0599-2
42. S. Stenlund, *Combinators, λ-Terms and Proof Theory*. 1972 ISBN 90-277-0305-1
43. M. Strauss, *Modern Physics and Its Philosophy*. Selected Paper in the Logic, History, and Philosophy of Science. 1972 ISBN 90-277-0230-6
44. M. Bunge, *Method, Model and Matter*. 1973 ISBN 90-277-0252-7
45. M. Bunge, *Philosophy of Physics*. 1973 ISBN 90-277-0253-5

46. A. A. Zinov'ev, *Foundations of the Logical Theory of Scientific Knowledge (Complex Logic)*. Revised and enlarged English edition with an appendix by G. A. Smirnov, E. A. Sidorenka, A. M. Fedina and L. A. Bobrova. [Boston Studies in the Philosophy of Science, Vol. IX] 1973 ISBN 90-277-0193-8; Pb 90-277-0324-8
47. L. Tondl, *Scientific Procedures*. A Contribution concerning the Methodological Problems of Scientific Concepts and Scientific Explanation. Translated from Czech by D. Short. Edited by R.S. Cohen and M.W. Wartofsky. [Boston Studies in the Philosophy of Science, Vol. X] 1973 ISBN 90-277-0147-4; Pb 90-277-0323-X
48. N. R. Hanson, *Constellations and Conjectures*. 1973 ISBN 90-277-0192-X
49. K. J. J. Hintikka, J. M. E. Moravcsik and P. Suppes (eds.), *Approaches to Natural Language*. 1973 ISBN 90-277-0220-9; Pb 90-277-0233-0
50. M. Bunge (ed.), *Exact Philosophy*. Problems, Tools and Goals. 1973 ISBN 90-277-0251-9
51. R. J. Bogdan and I. Niiniluoto (eds.), *Logic, Language and Probability*. 1973 ISBN 90-277-0312-4
52. G. Pearce and P. Maynard (eds.), *Conceptual Change*. 1973 ISBN 90-277-0287-X; Pb 90-277-0339-6
53. I. Niiniluoto and R. Tuomela, *Theoretical Concepts and Hypothetico-inductive Inference*. 1973 ISBN 90-277-0343-4
54. R. Fraissé, *Course of Mathematical Logic* – Volume 1: *Relation and Logical Formula*. Translated from French. 1973 ISBN 90-277-0268-3; Pb 90-277-0403-1 (For *Volume 2* see under No. 69).
55. A. Grünbaum, *Philosophical Problems of Space and Time*. Edited by R.S. Cohen and M.W. Wartofsky. 2nd enlarged ed. [Boston Studies in the Philosophy of Science, Vol. XII] 1973 ISBN 90-277-0357-4; Pb 90-277-0358-2
56. P. Suppes (ed.), *Space, Time and Geometry*. 1973 ISBN 90-277-0386-8; Pb 90-277-0442-2
57. H. Kelsen, *Essays in Legal and Moral Philosophy*. Selected and introduced by O. Weinberger. Translated from German by P. Heath. 1973 ISBN 90-277-0388-4
58. R. J. Seeger and R. S. Cohen (eds.), *Philosophical Foundations of Science*. [Boston Studies in the Philosophy of Science, Vol. XI] 1974 ISBN 90-277-0390-6; Pb 90-277-0376-0
59. R. S. Cohen and M. W. Wartofsky (eds.), *Logical and Epistemological Studies in Contemporary Physics*. [Boston Studies in the Philosophy of Science, Vol. XIII] 1973 ISBN 90-277-0391-4; Pb 90-277-0377-9
60. R. S. Cohen and M. W. Wartofsky (eds.), *Methodological and Historical Essays in the Natural and Social Sciences. Proceedings of the Boston Colloquium for the Philosophy of Science, 1969–1972*. [Boston Studies in the Philosophy of Science, Vol. XIV] 1974 ISBN 90-277-0392-2; Pb 90-277-0378-7
61. R. S. Cohen, J. J. Stachel and M. W. Wartofsky (eds.), *For Dirk Struik. Scientific, Historical and Political Essays*. [Boston Studies in the Philosophy of Science, Vol. XV] 1974 ISBN 90-277-0393-0; Pb 90-277-0379-5
62. K. Ajdukiewicz, *Pragmatic Logic*. Translated from Polish by O. Wojtasiewicz. 1974 ISBN 90-277-0326-4
63. S. Stenlund (ed.), *Logical Theory and Semantic Analysis*. Essays dedicated to Stig Kanger on His 50th Birthday. 1974 ISBN 90-277-0438-4
64. K. F. Schaffner and R. S. Cohen (eds.), *PSA 1972. Proceedings of the Third Biennial Meeting of the Philosophy of Science Association*. [Boston Studies in the Philosophy of Science, Vol. XX] 1974 ISBN 90-277-0408-2; Pb 90-277-0409-0

65. H. E. Kyburg, Jr., *The Logical Foundations of Statistical Inference*. 1974
ISBN 90-277-0330-2; Pb 90-277-0430-9
66. M. Grene, *The Understanding of Nature*. Essays in the Philosophy of Biology. [Boston Studies in the Philosophy of Science, Vol. XXIII] 1974
ISBN 90-277-0462-7; Pb 90-277-0463-5
67. J. M. Broekman, *Structuralism: Moscow, Prague, Paris*. Translated from German. 1974 ISBN 90-277-0478-3
68. N. Geschwind, *Selected Papers on Language and the Brain*. [Boston Studies in the Philosophy of Science, Vol. XVI] 1974 ISBN 90-277-0262-4; Pb 90-277-0263-2
69. R. Fraissé, *Course of Mathematical Logic* – Volume 2: *Model Theory*. Translated from French. 1974 ISBN 90-277-0269-1; Pb 90-277-0510-0
(For *Volume 1* see under No. 54)
70. A. Grzegorczyk, *An Outline of Mathematical Logic*. Fundamental Results and Notions explained with all Details. Translated from Polish. 1974
ISBN 90-277-0359-0; Pb 90-277-0447-3
71. F. von Kutschera, *Philosophy of Language*. 1975 ISBN 90-277-0591-7
72. J. Manninen and R. Tuomela (eds.), *Essays on Explanation and Understanding*. Studies in the Foundations of Humanities and Social Sciences. 1976
ISBN 90-277-0592-5
73. J. Hintikka (ed.), *Rudolf Carnap, Logical Empiricist*. Materials and Perspectives. 1975 ISBN 90-277-0583-6
74. M. Čapek (ed.), *The Concepts of Space and Time*. Their Structure and Their Development. [Boston Studies in the Philosophy of Science, Vol. XXII] 1976
ISBN 90-277-0355-8; Pb 90-277-0375-2
75. J. Hintikka and U. Remes, *The Method of Analysis*. Its Geometrical Origin and Its General Significance. [Boston Studies in the Philosophy of Science, Vol. XXV] 1974 ISBN 90-277-0532-1; Pb 90-277-0543-7
76. J. E. Murdoch and E. D. Sylla (eds.), *The Cultural Context of Medieval Learning*. [Boston Studies in the Philosophy of Science, Vol. XXVI] 1975
ISBN 90-277-0560-7; Pb 90-277-0587-9
77. S. Amsterdamski, *Between Experience and Metaphysics*. Philosophical Problems of the Evolution of Science. [Boston Studies in the Philosophy of Science, Vol. XXXV] 1975 ISBN 90-277-0568-2; Pb 90-277-0580-1
78. P. Suppes (ed.), *Logic and Probability in Quantum Mechanics*. 1976
ISBN 90-277-0570-4; Pb 90-277-1200-X
79. H. von Helmholtz: *Epistemological Writings. The Paul Hertz / Moritz Schlick Centenary Edition of 1921 with Notes and Commentary by the Editors*. Newly translated from German by M. F. Lowe. Edited, with an Introduction and Bibliography, by R. S. Cohen and Y. Elkana. [Boston Studies in the Philosophy of Science, Vol. XXXVII] 1975 ISBN 90-277-0290-X; Pb 90-277-0582-8
80. J. Agassi, *Science in Flux*. [Boston Studies in the Philosophy of Science, Vol. XXVIII] 1975 ISBN 90-277-0584-4; Pb 90-277-0612-2
81. S. G. Harding (ed.), *Can Theories Be Refuted?* Essays on the Duhem-Quine Thesis. 1976 ISBN 90-277-0629-8; Pb 90-277-0630-1
82. S. Nowak, *Methodology of Sociological Research*. General Problems. 1977
ISBN 90-277-0486-4
83. J. Piaget, J.-B. Grize, A. Szemińska and V. Bang, *Epistemology and Psychology of Functions*. Translated from French. 1977 ISBN 90-277-0804-5

84. M. Grene and E. Mendelsohn (eds.), *Topics in the Philosophy of Biology.* [Boston Studies in the Philosophy of Science, Vol. XXVII] 1976
ISBN 90-277-0595-X; Pb 90-277-0596-8
85. E. Fischbein, *The Intuitive Sources of Probabilistic Thinking in Children.* 1975
ISBN 90-277-0626-3; Pb 90-277-1190-9
86. E. W. Adams, *The Logic of Conditionals.* An Application of Probability to Deductive Logic. 1975 ISBN 90-277-0631-X
87. M. Przełęcki and R. Wójcicki (eds.), *Twenty-Five Years of Logical Methodology in Poland.* Translated from Polish. 1976 ISBN 90-277-0601-8
88. J. Topolski, *The Methodology of History.* Translated from Polish by O. Wojtasiewicz. 1976 ISBN 90-277-0550-X
89. A. Kasher (ed.), *Language in Focus: Foundations, Methods and Systems.* Essays dedicated to Yehoshua Bar-Hillel. [Boston Studies in the Philosophy of Science, Vol. XLIII] 1976 ISBN 90-277-0644-1; Pb 90-277-0645-X
90. J. Hintikka, *The Intentions of Intentionality and Other New Models for Modalities.* 1975 ISBN 90-277-0633-6; Pb 90-277-0634-4
91. W. Stegmüller, *Collected Papers on Epistemology, Philosophy of Science and History of Philosophy.* 2 Volumes. 1977 Set ISBN 90-277-0767-7
92. D. M. Gabbay, *Investigations in Modal and Tense Logics with Applications to Problems in Philosophy and Linguistics.* 1976 ISBN 90-277-0656-5
93. R. J. Bogdan, *Local Induction.* 1976 ISBN 90-277-0649-2
94. S. Nowak, *Understanding and Prediction.* Essays in the Methodology of Social and Behavioral Theories. 1976 ISBN 90-277-0558-5; Pb 90-277-1199-2
95. P. Mittelstaedt, *Philosophical Problems of Modern Physics.* [Boston Studies in the Philosophy of Science, Vol. XVIII] 1976 ISBN 90-277-0285-3; Pb 90-277-0506-2
96. G. Holton and W. A. Blanpied (eds.), *Science and Its Public: The Changing Relationship.* [Boston Studies in the Philosophy of Science, Vol. XXXIII] 1976
ISBN 90-277-0657-3; Pb 90-277-0658-1
97. M. Brand and D. Walton (eds.), *Action Theory.* 1976 ISBN 90-277-0671-9
98. P. Gochet, *Outline of a Nominalist Theory of Propositions.* An Essay in the Theory of Meaning and in the Philosophy of Logic. 1980 ISBN 90-277-1031-7
99. R. S. Cohen, P. K. Feyerabend, and M. W. Wartofsky (eds.), *Essays in Memory of Imre Lakatos.* [Boston Studies in the Philosophy of Science, Vol. XXXIX] 1976
ISBN 90-277-0654-9; Pb 90-277-0655-7
100. R. S. Cohen and J. J. Stachel (eds.), *Selected Papers of Léon Rosenfield.* [Boston Studies in the Philosophy of Science, Vol. XXI] 1979
ISBN 90-277-0651-4; Pb 90-277-0652-2
101. R. S. Cohen, C. A. Hooker, A. C. Michalos and J. W. van Evra (eds.), *PSA 1974. Proceedings of the 1974 Biennial Meeting of the Philosophy of Science Association.* [Boston Studies in the Philosophy of Science, Vol. XXXII] 1976
ISBN 90-277-0647-6; Pb 90-277-0648-4
102. Y. Fried and J. Agassi, *Paranoia.* A Study in Diagnosis. [Boston Studies in the Philosophy of Science, Vol. L] 1976 ISBN 90-277-0704-9; Pb 90-277-0705-7
103. M. Przełęcki, K. Szaniawski and R. Wójcicki (eds.), *Formal Methods in the Methodology of Empirical Sciences.* 1976 ISBN 90-277-0698-0
104. J. M. Vickers, *Belief and Probability.* 1976 ISBN 90-277-0744-8
105. K. H. Wolff, *Surrender and Catch.* Experience and Inquiry Today. [Boston Studies in the Philosophy of Science, Vol. LI] 1976
ISBN 90-277-0758-8; Pb 90-277-0765-0

106. K. Kosík, *Dialectics of the Concrete.* A Study on Problems of Man and World. [Boston Studies in the Philosophy of Science, Vol. LII] 1976
ISBN 90-277-0761-8; Pb 90-277-0764-2
107. N. Goodman, *The Structure of Appearance.* 3rd ed. with an Introduction by G. Hellman. [Boston Studies in the Philosophy of Science, Vol. LIII] 1977
ISBN 90-277-0773-1; Pb 90-277-0774-X
108. K. Ajdukiewicz, *The Scientific World-Perspective and Other Essays, 1931-1963.* Translated from Polish. Edited and with an Introduction by J. Giedymin. 1978
ISBN 90-277-0527-5
109. R. L. Causey, *Unity of Science.* 1977 ISBN 90-277-0779-0
110. R. E. Grandy, *Advanced Logic for Applications.* 1977 ISBN 90-277-0781-2
111. R. P. McArthur, *Tense Logic.* 1976 ISBN 90-277-0697-2
112. L. Lindahl, *Position and Change.* A Study in Law and Logic. Translated from Swedish by P. Needham. 1977 ISBN 90-277-0787-1
113. R. Tuomela, *Dispositions.* 1978 ISBN 90-277-0810-X
114. H. A. Simon, *Models of Discovery and Other Topics in the Methods of Science.* [Boston Studies in the Philosophy of Science, Vol. LIV] 1977
ISBN 90-277-0812-6; Pb 90-277-0858-4
115. R. D. Rosenkrantz, *Inference, Method and Decision.* Towards a Bayesian Philosophy of Science. 1977 ISBN 90-277-0817-7; Pb 90-277-0818-5
116. R. Tuomela, *Human Action and Its Explanation.* A Study on the Philosophical Foundations of Psychology. 1977 ISBN 90-277-0824-X
117. M. Lazerowitz, *The Language of Philosophy.* Freud and Wittgenstein. [Boston Studies in the Philosophy of Science, Vol. LV] 1977
ISBN 90-277-0826-6; Pb 90-277-0862-2
118. Not published
119. J. Pelc (ed.), *Semiotics in Poland, 1894–1969.* Translated from Polish. 1979
ISBN 90-277-0811-8
120. I. Pörn, *Action Theory and Social Science.* Some Formal Models. 1977
ISBN 90-277-0846-0
121. J. Margolis, *Persons and Mind.* The Prospects of Nonreductive Materialism. [Boston Studies in the Philosophy of Science, Vol. LVII] 1977
ISBN 90-277-0854-1; Pb 90-277-0863-0
122. J. Hintikka, I. Niiniluoto, and E. Saarinen (eds.), *Essays on Mathematical and Philosophical Logic.* 1979 ISBN 90-277-0879-7
123. T. A. F. Kuipers, *Studies in Inductive Probability and Rational Expectation.* 1978
ISBN 90-277-0882-7
124. E. Saarinen, R. Hilpinen, I. Niiniluoto and M. P. Hintikka (eds.), *Essays in Honour of Jaakko Hintikka on the Occasion of His 50th Birthday.* 1979
ISBN 90-277-0916-5
125. G. Radnitzky and G. Andersson (eds.), *Progress and Rationality in Science.* [Boston Studies in the Philosophy of Science, Vol. LVIII] 1978
ISBN 90-277-0921-1; Pb 90-277-0922-X
126. P. Mittelstaedt, *Quantum Logic.* 1978 ISBN 90-277-0925-4
127. K. A. Bowen, *Model Theory for Modal Logic.* Kripke Models for Modal Predicate Calculi. 1979 ISBN 90-277-0929-7
128. H. A. Bursen, *Dismantling the Memory Machine.* A Philosophical Investigation of Machine Theories of Memory. 1978 ISBN 90-277-0933-5

129. M. W. Wartofsky, *Models.* Representation and the Scientific Understanding. [Boston Studies in the Philosophy of Science, Vol. XLVIII] 1979
ISBN 90-277-0736-7; Pb 90-277-0947-5
130. D. Ihde, *Technics and Praxis.* A Philosophy of Technology. [Boston Studies in the Philosophy of Science, Vol. XXIV] 1979 ISBN 90-277-0953-X; Pb 90-277-0954-8
131. J. J. Wiatr (ed.), *Polish Essays in the Methodology of the Social Sciences.* [Boston Studies in the Philosophy of Science, Vol. XXIX] 1979
ISBN 90-277-0723-5; Pb 90-277-0956-4
132. W. C. Salmon (ed.), *Hans Reichenbach: Logical Empiricist.* 1979
ISBN 90-277-0958-0
133. P. Bieri, R.-P. Horstmann and L. Krüger (eds.), *Transcendental Arguments in Science.* Essays in Epistemology. 1979 ISBN 90-277-0963-7; Pb 90-277-0964-5
134. M. Marković and G. Petrović (eds.), *Praxis.* Yugoslav Essays in the Philosophy and Methodology of the Social Sciences. [Boston Studies in the Philosophy of Science, Vol. XXXVI] 1979 ISBN 90-277-0727-8; Pb 90-277-0968-8
135. R. Wójcicki, *Topics in the Formal Methodology of Empirical Sciences.* Translated from Polish. 1979 ISBN 90-277-1004-X
136. G. Radnitzky and G. Andersson (eds.), *The Structure and Development of Science.* [Boston Studies in the Philosophy of Science, Vol. LIX] 1979
ISBN 90-277-0994-7; Pb 90-277-0995-5
137. J. C. Webb, *Mechanism, Mentalism and Metamathematics.* An Essay on Finitism. 1980 ISBN 90-277-1046-5
138. D. F. Gustafson and B. L. Tapscott (eds.), *Body, Mind and Method.* Essays in Honor of Virgil C. Aldrich. 1979 ISBN 90-277-1013-9
139. L. Nowak, *The Structure of Idealization.* Towards a Systematic Interpretation of the Marxian Idea of Science. 1980 ISBN 90-277-1014-7
140. C. Perelman, *The New Rhetoric and the Humanities.* Essays on Rhetoric and Its Applications. Translated from French and German. With an Introduction by H. Zyskind. 1979 ISBN 90-277-1018-X; Pb 90-277-1019-8
141. W. Rabinowicz, *Universalizability.* A Study in Morals and Metaphysics. 1979
ISBN 90-277-1020-2
142. C. Perelman, *Justice, Law and Argument.* Essays on Moral and Legal Reasoning. Translated from French and German. With an Introduction by H.J. Berman. 1980
ISBN 90-277-1089-9; Pb 90-277-1090-2
143. S. Kanger and S. Öhman (eds.), *Philosophy and Grammar.* Papers on the Occasion of the Quincentennial of Uppsala University. 1981 ISBN 90-277-1091-0
144. T. Pawlowski, *Concept Formation in the Humanities and the Social Sciences.* 1980
ISBN 90-277-1096-1
145. J. Hintikka, D. Gruender and E. Agazzi (eds.), *Theory Change, Ancient Axiomatics and Galileo's Methodology.* Proceedings of the 1978 Pisa Conference on the History and Philosophy of Science, Volume I. 1981 ISBN 90-277-1126-7
146. J. Hintikka, D. Gruender and E. Agazzi (eds.), *Probabilistic Thinking, Thermodynamics, and the Interaction of the History and Philosophy of Science.* Proceedings of the 1978 Pisa Conference on the History and Philosophy of Science, Volume II. 1981 ISBN 90-277-1127-5
147. U. Mönnich (ed.), *Aspects of Philosophical Logic.* Some Logical Forays into Central Notions of Linguistics and Philosophy. 1981 ISBN 90-277-1201-8
148. D. M. Gabbay, *Semantical Investigations in Heyting's Intuitionistic Logic.* 1981
ISBN 90-277-1202-6

149. E. Agazzi (ed.), *Modern Logic – A Survey*. Historical, Philosophical, and Mathematical Aspects of Modern Logic and Its Applications. 1981 ISBN 90-277-1137-2
150. A. F. Parker-Rhodes, *The Theory of Indistinguishables*. A Search for Explanatory Principles below the Level of Physics. 1981 ISBN 90-277-1214-X
151. J. C. Pitt, *Pictures, Images, and Conceptual Change*. An Analysis of Wilfrid Sellars' Philosophy of Science. 1981 ISBN 90-277-1276-X; Pb 90-277-1277-8
152. R. Hilpinen (ed.), *New Studies in Deontic Logic*. Norms, Actions, and the Foundations of Ethics. 1981 ISBN 90-277-1278-6; Pb 90-277-1346-4
153. C. Dilworth, *Scientific Progress*. A Study Concerning the Nature of the Relation between Successive Scientific Theories. 2nd, rev. and augmented ed., 1986 ISBN 90-277-2215-3; Pb 90-277-2216-1
154. D. Woodruff Smith and R. McIntyre, *Husserl and Intentionality*. A Study of Mind, Meaning, and Language. 1982 ISBN 90-277-1392-8; Pb 90-277-1730-3
155. R. J. Nelson, *The Logic of Mind*. 2nd. ed., 1989 ISBN 90-277-2819-4; Pb 90-277-2822-4
156. J. F. A. K. van Benthem, *The Logic of Time*. A Model-Theoretic Investigation into the Varieties of Temporal Ontology, and Temporal Discourse. 1983; 2nd ed., 1991 ISBN 0-7923-1081-0
157. R. Swinburne (ed.), *Space, Time and Causality*. 1983 ISBN 90-277-1437-1
158. E. T. Jaynes, *Papers on Probability, Statistics and Statistical Physics*. Ed. by R. D. Rozenkrantz. 1983 ISBN 90-277-1448-7; Pb (1989) 0-7923-0213-3
159. T. Chapman, *Time: A Philosophical Analysis*. 1982 ISBN 90-277-1465-7
160. E. N. Zalta, *Abstract Objects*. An Introduction to Axiomatic Metaphysics. 1983 ISBN 90-277-1474-6
161. S. Harding and M. B. Hintikka (eds.), *Discovering Reality*. Feminist Perspectives on Epistemology, Metaphysics, Methodology, and Philosophy of Science. 1983 ISBN 90-277-1496-7; Pb 90-277-1538-6
162. M. A. Stewart (ed.), *Law, Morality and Rights*. 1983 ISBN 90-277-1519-X
163. D. Mayr and G. Süssmann (eds.), *Space, Time, and Mechanics*. Basic Structures of a Physical Theory. 1983 ISBN 90-277-1525-4
164. D. Gabbay and F. Guenthner (eds.), *Handbook of Philosophical Logic*. Vol. I: Elements of Classical Logic. 1983 ISBN 90-277-1542-4
165. D. Gabbay and F. Guenthner (eds.), *Handbook of Philosophical Logic*. Vol. II: Extensions of Classical Logic. 1984 ISBN 90-277-1604-8
166. D. Gabbay and F. Guenthner (eds.), *Handbook of Philosophical Logic*. Vol. III: Alternative to Classical Logic. 1986 ISBN 90-277-1605-6
167. D. Gabbay and F. Guenthner (eds.), *Handbook of Philosophical Logic*. Vol. IV: Topics in the Philosophy of Language. 1989 ISBN 90-277-1606-4
168. A. J. I. Jones, *Communication and Meaning*. An Essay in Applied Modal Logic. 1983 ISBN 90-277-1543-2
169. M. Fitting, *Proof Methods for Modal and Intuitionistic Logics*. 1983 ISBN 90-277-1573-4
170. J. Margolis, *Culture and Cultural Entities*. Toward a New Unity of Science. 1984 ISBN 90-277-1574-2
171. R. Tuomela, *A Theory of Social Action*. 1984 ISBN 90-277-1703-6
172. J. J. E. Gracia, E. Rabossi, E. Villanueva and M. Dascal (eds.), *Philosophical Analysis in Latin America*. 1984 ISBN 90-277-1749-4
173. P. Ziff, *Epistemic Analysis*. A Coherence Theory of Knowledge. 1984 ISBN 90-277-1751-7

174. P. Ziff, *Antiaesthetics*. An Appreciation of the Cow with the Subtile Nose. 1984 ISBN 90-277-1773-7
175. W. Balzer, D. A. Pearce, and H.-J. Schmidt (eds.), *Reduction in Science*. Structure, Examples, Philosophical Problems. 1984 ISBN 90-277-1811-3
176. A. Peczenik, L. Lindahl and B. van Roermund (eds.), *Theory of Legal Science*. Proceedings of the Conference on Legal Theory and Philosophy of Science (Lund, Sweden, December 1983). 1984 ISBN 90-277-1834-2
177. I. Niiniluoto, *Is Science Progressive?* 1984 ISBN 90-277-1835-0
178. B. K. Matilal and J. L. Shaw (eds.), *Analytical Philosophy in Comparative Perspective*. Exploratory Essays in Current Theories and Classical Indian Theories of Meaning and Reference. 1985 ISBN 90-277-1870-9
179. P. Kroes, *Time: Its Structure and Role in Physical Theories*. 1985 ISBN 90-277-1894-6
180. J. H. Fetzer, *Sociobiology and Epistemology*. 1985 ISBN 90-277-2005-3; Pb 90-277-2006-1
181. L. Haaparanta and J. Hintikka (eds.), *Frege Synthesized*. Essays on the Philosophical and Foundational Work of Gottlob Frege. 1986 ISBN 90-277-2126-2
182. M. Detlefsen, *Hilbert's Program*. An Essay on Mathematical Instrumentalism. 1986 ISBN 90-277-2151-3
183. J. L. Golden and J. J. Pilotta (eds.), *Practical Reasoning in Human Affairs*. Studies in Honor of Chaim Perelman. 1986 ISBN 90-277-2255-2
184. H. Zandvoort, *Models of Scientific Development and the Case of Nuclear Magnetic Resonance*. 1986 ISBN 90-277-2351-6
185. I. Niiniluoto, *Truthlikeness*. 1987 ISBN 90-277-2354-0
186. W. Balzer, C. U. Moulines and J. D. Sneed, *An Architectonic for Science*. The Structuralist Program. 1987 ISBN 90-277-2403-2
187. D. Pearce, *Roads to Commensurability*. 1987 ISBN 90-277-2414-8
188. L. M. Vaina (ed.), *Matters of Intelligence*. Conceptual Structures in Cognitive Neuroscience. 1987 ISBN 90-277-2460-1
189. H. Siegel, *Relativism Refuted*. A Critique of Contemporary Epistemological Relativism. 1987 ISBN 90-277-2469-5
190. W. Callebaut and R. Pinxten, *Evolutionary Epistemology*. A Multiparadigm Program, with a Complete Evolutionary Epistemology Bibliograph. 1987 ISBN 90-277-2582-9
191. J. Kmita, *Problems in Historical Epistemology*. 1988 ISBN 90-277-2199-8
192. J. H. Fetzer (ed.), *Probability and Causality*. Essays in Honor of Wesley C. Salmon, with an Annotated Bibliography. 1988 ISBN 90-277-2607-8; Pb 1-5560-8052-2
193. A. Donovan, L. Laudan and R. Laudan (eds.), *Scrutinizing Science*. Empirical Studies of Scientific Change. 1988 ISBN 90-277-2608-6
194. H.R. Otto and J.A. Tuedio (eds.), *Perspectives on Mind*. 1988 ISBN 90-277-2640-X
195. D. Batens and J.P. van Bendegem (eds.), *Theory and Experiment*. Recent Insights and New Perspectives on Their Relation. 1988 ISBN 90-277-2645-0
196. J. Österberg, *Self and Others*. A Study of Ethical Egoism. 1988 ISBN 90-277-2648-5
197. D.H. Helman (ed.), *Analogical Reasoning*. Perspectives of Artificial Intelligence, Cognitive Science, and Philosophy. 1988 ISBN 90-277-2711-2
198. J. Wolenski, *Logic and Philosophy in the Lvov-Warsaw School*. 1989 ISBN 90-277-2749-X

199. R. Wójcicki, *Theory of Logical Calculi*. Basic Theory of Consequence Operations. 1988 ISBN 90-277-2785-6
200. J. Hintikka and M.B. Hintikka, *The Logic of Epistemology and the Epistemology of Logic*. Selected Essays. 1989 ISBN 0-7923-0040-8; Pb 0-7923-0041-6
201. E. Agazzi (ed.), *Probability in the Sciences*. 1988 ISBN 90-277-2808-9
202. M. Meyer (ed.), *From Metaphysics to Rhetoric*. 1989 ISBN 90-277-2814-3
203. R.L. Tieszen, *Mathematical Intuition*. Phenomenology and Mathematical Knowledge. 1989 ISBN 0-7923-0131-5
204. A. Melnick, *Space, Time, and Thought in Kant*. 1989 ISBN 0-7923-0135-8
205. D.W. Smith, *The Circle of Acquaintance*. Perception, Consciousness, and Empathy. 1989 ISBN 0-7923-0252-4
206. M.H. Salmon (ed.), *The Philosophy of Logical Mechanism*. Essays in Honor of Arthur W. Burks. With his Responses, and with a Bibliography of Burk's Work. 1990 ISBN 0-7923-0325-3
207. M. Kusch, *Language as Calculus vs. Language as Universal Medium*. A Study in Husserl, Heidegger, and Gadamer. 1989 ISBN 0-7923-0333-4
208. T.C. Meyering, *Historical Roots of Cognitive Science*. The Rise of a Cognitive Theory of Perception from Antiquity to the Nineteenth Century. 1989 ISBN 0-7923-0349-0
209. P. Kosso, *Observability and Observation in Physical Science*. 1989 ISBN 0-7923-0389-X
210. J. Kmita, *Essays on the Theory of Scientific Cognition*. 1990 ISBN 0-7923-0441-1
211. W. Sieg (ed.), *Acting and Reflecting*. The Interdisciplinary Turn in Philosophy. 1990 ISBN 0-7923-0512-4
212. J. Karpiński, *Causality in Sociological Research*. 1990 ISBN 0-7923-0546-9
213. H.A. Lewis (ed.), *Peter Geach: Philosophical Encounters*. 1991 ISBN 0-7923-0823-9
214. M. Ter Hark, *Beyond the Inner and the Outer*. Wittgenstein's Philosophy of Psychology. 1990 ISBN 0-7923-0850-6
215. M. Gosselin, *Nominalism and Contemporary Nominalism*. Ontological and Epistemological Implications of the Work of W.V.O. Quine and of N. Goodman. 1990 ISBN 0-7923-0904-9
216. J.H. Fetzer, D. Shatz and G. Schlesinger (eds.), *Definitions and Definability*. Philosophical Perspectives. 1991 ISBN 0-7923-1046-2
217. E. Agazzi and A. Cordero (eds.), *Philosophy and the Origin and Evolution of the Universe*. 1991 ISBN 0-7923-1322-4
218. M. Kusch, *Foucault's Strata and Fields*. An Investigation into Archaeological and Genealogical Science Studies. 1991 ISBN 0-7923-1462-X
219. C.J. Posy, *Kant's Philosophy of Mathematics*. Modern Essays. 1992 ISBN 0-7923-1495-6
220. G. Van de Vijver, *New Perspectives on Cybernetics*. Self-Organization, Autonomy and Connectionism. 1992 ISBN 0-7923-1519-7

Previous volumes are still available.

KLUWER ACADEMIC PUBLISHERS – DORDRECHT / BOSTON / LONDON

GPSR Compliance

The European Union's (EU) General Product Safety Regulation (GPSR) is a set of rules that requires consumer products to be safe and our obligations to ensure this.

If you have any concerns about our products, you can contact us on ProductSafety@springernature.com

In case Publisher is established outside the EU, the EU authorized representative is:

Springer Nature Customer Service Center GmbH
Europaplatz 3
69115 Heidelberg, Germany

Batch number: 10823811

Printed by Printforce, the Netherlands